Cambridge Elements

Elements in Quantitative and Computational Methods
for the Social Sciences
edited by
R. Michael Alvarez
California Institute of Technology
Nathaniel Beck
New York University
Betsy Sinclair
Washington University in St. Louis

BAYESIAN SOCIAL SCIENCE STATISTICS

From the Very Beginning

Jeff Gill
American University
Le Bao
Georgetown University

Shaftesbury Road, Cambridge CB2 8EA, United Kingdom

One Liberty Plaza, 20th Floor, New York, NY 10006, USA

477 Williamstown Road, Port Melbourne, VIC 3207, Australia

314–321, 3rd Floor, Plot 3, Splendor Forum, Jasola District Centre, New Delhi – 110025, India

103 Penang Road, #05–06/07, Visioncrest Commercial, Singapore 238467

Cambridge University Press is part of Cambridge University Press & Assessment, a department of the University of Cambridge.

We share the University's mission to contribute to society through the pursuit of education, learning and research at the highest international levels of excellence.

www.cambridge.org
Information on this title: www.cambridge.org/9781009494694

DOI: 10.1017/9781009341189

When citing this work, please include a reference to the DOI 10.1017/9781009341189

First published 2024

A catalogue record for this publication is available from the British Library.

ISBN 978-1-009-49469-4 Hardback
ISBN 978-1-009-34119-6 Paperback
ISSN 2398-4023 (online)
ISSN 2514-3794 (print)

Additional resources for this publication at www.cambridge.org/Gill-Bao

Bayesian Social Science Statistics

From the Very Beginning

Elements in Quantitative and Computational Methods for the Social Sciences

DOI: 10.1017/9781009341189
First published online: May 2024

Jeff Gill
American University

Le Bao
Georgetown University

Author for correspondence: Jeff Gill, jgill@american.edu

Abstract: In this Element, the authors introduce Bayesian probability and inference for social science students and practitioners starting from the absolute beginning and walk readers steadily through the Element. No previous knowledge is required other than that in a basic statistics course. At the end of the process, readers will understand the core tenets of Bayesian theory and practice in a way that enables them to specify, implement, and understand models using practical social science data. Sections will cover theoretical principles and real-world applications that provide motivation and intuition. Because Bayesian methods are intricately tied to software, code in both R and Python is provided throughout.

Keywords: Bayesian statistics, probability, prior distribution, posterior distribution, likelihood function

ISBNs: 9781009494694 (HB), 9781009341196 (PB), 9781009341189 (OC)
ISSNs: 2398-4023 (online), 2514-3794 (print)

Contents

1 Introduction: The Purpose and Scope of This Element

The Bayesian philosophy about statistical inference is much older than the better-established traditional Frequentist/Likelihoodist paradigm, starting with Bayes (1763) and considered by Laplace, Gauss, and others. Scholars such as Jeffreys, Zellner, Savage, de Finetti, and Lindley reactivated interest in Bayesian methods in the middle of the last century in response to observed deficiencies in classical techniques. Unfortunately many of the specifications developed by these early Bayesians, while superior in theoretical foundation, led to mathematically intractable forms. This problem has been famously solved in recent years by a revolution in statistical computing techniques. Yet regrettably these remarkable developments have not permeated all data-analytic academic fields including the social and behavioral sciences in particular. This may be because of a lack of effective introductory material.

The Bayesian paradigm is ideally suited to the type of data analysis performed by social scientists because it recognizes the mobility of population parameters, incorporates prior knowledge that researchers possess, and updates estimates as new data are observed. Because most empirical work in the social sciences is observational rather than experimental, subjects are rarely cooperative, and systematic effects are often more elusive than in other fields, the Bayesian approach to modeling uncertainty in parameter estimates provides a more robust and realistic picture of the data generating process.

Most people would be surprised to hear that there are different philosophical views within statistics. In fact, this led to a major split in the discipline through most of the twentieth century. A big reason for this schism was that most of the giants of the adolescent age of statistics in the late nineteenth and early twentieth century were openly hostile towards the Bayesian paradigm. Another reason was (notice the past tense) that computing resources were not available in this period for producing useful Bayesian results for the types of models that statisticians and others wanted to specify. Neither of these two issues are important now. It is still useful to understand how the typology of statistics still permeates written and spoken discussions today.

The first generation of modern statisticians were Frequentists from the idea of performing some experiment or test multiple times using a long stream of independent observations from the same data generating source. If a frequentist wanted to determine the probability of heads when presented with a new coin, the procedure would be to flip it into a sand pit hundreds or thousands of times and record the long-term results. This comes from the classic Neyman/Pearson/Wald setup in the late nineteenth century where the orthodox view is sampling is infinite (or as long as desired) and decision rules can be sharp.

They also mostly studied physical phenomenon and therefore viewed many target population parameters as fixed by nature where variation occurred from less than perfectly accurate instruments and observers. This idea of replicated tests is a key part of frequentist statistics and led to classic frequentist procedures such as the $1 - \alpha$ confidence interval: an interval that over $100(1 - \alpha)\%$ of replications contains the true value of the parameter on average. To this day the dichotomy between pure frequentism and the regular practice of social science statistics makes this definition confusing to both students and practitioners since it is literally built on the idea of repeating the exact same experiment multiple times, which few of us are fortunate enough to be able to do. Since variability to a frequentist rests in the data and the parameters are assumed fixed points by nature, the key probabilistic quantity of interest is the probability of some function of the data given some hypothesis: $p(f(\text{data})|H_A)$. The objective is a point estimate and the associated variance of this estimate that informs a formal procedure, setting some α level set in advance in a distributional test. In this setup H_A is accompanied by a *complementary* hypothesis H_B, meaning that the state of the world is explained by either. This sets up clear definition and calculation of Type I and Type II errors. The frequentist accepts H_A if $p(f(\text{data})|H_A) < \alpha$ and accepts H_B if $p(f(\text{data})|H_B) \geq \alpha$. That sentence should feel uncomfortable to every practicing empirical social scientist based on the use of the word "accept." This is the critical difference between actual frequentists and others who do not have a perceived infinite flow of independent identically distributed data and sharp complementary hypotheses. Nearly all social scientists have limited, often one-off, datasets that are contextual in time and space and cannot be replicated 19 more times to achieve a true 0.95 confidence interval. This why nearly everyone we know in the social sciences who calls themselves a frequentist is not a frequentist.

Why were the leading figures of early statistics openly hostile to the Bayesian approach? Fisher in particular objected to the uniform distribution as a choice of prior distribution, and most of them disliked the idea of inverting the order of conditional probability with Bayes' Law, which will be discussed at length in the next section. But there were other fundamental philosophical differences that ran counter to frequentist thinking. From the Bayes/Laplace/de Finetti tradition, all unknown quantities are treated probabilistically by assigning them a distribution or probability value, and the state of the world can always be updated by conditioning on new information as it arrives. In the absence of an unending stream of data, the data that do arrive are treated as observed and fixed by the limited sampling process. The most caustic arguments, though, centered around the interpretation of probability. To a Bayesian probability is not from a frequent, long-run, series of experiments but instead based on the "degree of belief" given necessarily limited information before or after a specific dataset arrives. This has been called a "subjective" interpretation of probability but this is a poor description since what the Bayesian interpretation really is based on is what does the currently existing evidence imply about an unknown parameter, say θ, using probability as the mechanism of description. So the key quantity of interest is $p(\theta|f(\text{data}))$, which is a distribution. Here $f(\text{data})$ is any applicable function of the data from a model or other treatment. Therefore instead of resorting to mechanical hypothesis testing a Bayesian can ask questions such as what is the probability that θ is greater than zero? Or what is the probability that a specific treatment changes the status of the treated group over the control group? Bayesians construct these inferential statements by taking a prior probability statement about the quantity of interest $p(\theta)$ and conditioning it on a new set of data to produce an updated version, $p(\theta|f(\text{data}))$, providing new knowledge if the data are informative.

The overall history of Bayesian statistics as described earlier is an almost soap opera like tale of scientific sociology. The Reverend Thomas Bayes died in 1761 without publishing his article "An Essay towards Solving a Problem in the Doctrine of Chances," which was then submitted on his behalf by his friend Richard Price (they are both buried in the small Bunhill Cemetery in central London, which you can visit). Some have conjectured that Bayes did not believe in or did not have full confidence in this work, therefore making Bayes not "Bayesian." This idea was later revisited in the late nineteenth through the middle twentieth centuries in particular by scholars such as Jeffreys, de Finetti, Good, Savage, Lindley, and Zellner. Their work was difficult because many of the giants of the time in statistics such as Pearson (Karl), Neyman and Pearson (Egon), Wald, and, of course, Fisher were openly hostile to the idea of Bayesian inference, particularly obtaining $p(\theta|f(\text{data}))$ using uniformly distributed prior distributions. Fascinating accounts can be found in Stigler (1982), Stigler (1983), and Dale (2012). The non-Bayesian charges included: priors are always assigned subjectively, inverting likelihood functions is illogical, and it is easy to specify Bayesian models such that unknown parameter estimation is difficult or impossible. The latter insult was the only one that had an element of truth to it, but this was solved in 1990 through the introduction of Markov chain Monte Carlo to use computational power rather than brute analytical derivation. We now exist in an era where there are no intellectual or analytical impediments to specifying and estimating Bayesian models.

This discussion here so far has left out the bulk of as-practiced social science statistics, which uses limited data collection resources to find evidence for an effect through statistics calculated once in space and time. This is generally from the central challenge in studying humans socially, politically, or biomedically: we are not fixed parameters of nature, and groups of humans interacting with each other are even less so. So because of the instability of human behavior, datasets tend to be limited in generality, and contextual to specific phenomenon being studied at that moment. It would be terrific to go back and get substantially more survey data relevant to the 2016 US presidential election to understand why predictions were so wrong, but we cannot do that even with unlimited funding. Armed with a single point estimate and uncertainty around it for some phenomenon of interest (maximum likelihood estimates as described in Section 3), social scientists are clearly closer to the Bayesian paradigm. In fact, all such models are special cases of Bayesian models with a uniform prior distribution (Section 4), and as the data size gets very large they are identical for any nonpathological choice of the prior distribution (also Section 4). So in fact the overwhelming majority of empirical social scientists are actually just Bayesians who do not yet know it. Hopefully this discussion is motivation to read on.

It is important to provide some pedagogical notes at this beginning point in the Bayesian journey provided here. First, one cannot do meaningful Bayesian inference in the social sciences or elsewhere without some *calculus* operations. This is perhaps one of the traditional impediments for wide acceptance and use of Bayesian statistics. So there will be some basic calculus in the statistical theory presented in a very gentle way here. It will be as direct and simple as possible with an emphasis on a general understanding of the procedures rather than detailed mathematical expositions. In the twenty-first century practicing empirical social scientists, data scientists, and others performing daily data analysis tasks do not need to do routine calculus calculations because almost invariably the computer does it for them. So the key task is to understand what these operations do rather than master the specific mathematical processes.

There are two primary calculus tools: *differentiation* and *integration*. Differentiation takes a function and determines what is the instantaneous rate of change for that function at a specific point. If you are driving a car you have some velocity but in the course of traveling you have changes in that velocity. For instance, after the light turns green you hit the gas and are not yet going very fast but your instantaneous rate of change is very high and positive because you are speeding up. Conversely, if you are cruising at a relatively steady state on the highway your instantaneous rate of change is very low even at high speed. Then as you see a traffic light or obstacle ahead you start to slow down using the break pedal and your instantaneous rate of change becomes large and negative. We often use derivatives to make a statement like this on functions to find points of interest like where this instantaneous rate of change is zero, which is a local maxima or minima of a curvilinear form. Integrals are instead about space. They can measure how much area is there under a curve to the x-axis between two points on that x-axis, like a slice of the curve above zero. This is very useful when dealing with probability functions and general Bayesian inference as we will shortly find out. The key point is to not worry so much about the notation, which some people find odd or intimidating looking, but to think broadly about what is being accomplished with these operations.

Second, we strongly emphasize using computer simulation to produce desired results rather than analytical calculations, which are of course traditional in twentieth-century Bayesian statistics. It turns out that nearly all of the quantities of interest can be produced faster and with less effort by requiring a machine to do repetitive work. This theme is also central to progress in Bayesian statistics that occurred right at the end of the twentieth century and led to enormous leaps in our ability to specify and estimate Bayesian models of interest. We will repeatedly make use of the simulation idea that quantities of interest, like estimated parameters, can be effectively described by generating many samples from the relevant distribution and then treating these values like data and doing simple summaries. This idea goes back to the dawn of computing and is vitally important in all areas of statistics in the twenty-first century.

All code and data used throughout this element are stored in the GitHub repository: https://github.com/jgill22/Bayesian.Social.Science.Statistics, and can also be conveniently executed online through the Code Ocean capsule located at: https://codeocean.com/capsule/8772484.

2 Basic Probability Principles and Bayes Law

Knowing basic probability theory is a requirement for understanding statistics, and even more so Bayesian statistics. Therefore in this section we introduce the required fundamentals and interested readers can consult Gill (2014) for more details. Probability is nothing more than a standard way to describe uncertainty. When unknown or future events are assigned a small positive number near zero we think of them as being very unlikely to occur. Conversely when these events are assigned a number less than but near one we think of them as being very likely to occur. When we perform the experiment of flipping a fair coin then each event/outcome, heads versus tails, is equally likely to happen so we assign them both the probability of occurrence of 0.5. Implied by this brief discussion is a set of rules that map the occurrence of events to their probability of happening. Start with a set of nonoverlapping discrete events, $A_1, A_2, \ldots, A_k$ that make up a sample space, $\mathcal{S}$, which is the collection of all things that can happen for this experiment. The probabilities associated with these events are denoted $p(A_1), p(A_2), \ldots, p(A_k)$, which are just functions in the spirit of $f(x) = x^2$ except we use $p()$ instead of $f()$ as reminder that these are *probability* functions. Returning to the example of flipping a fair coin we have $A_1 = heads$, and $A_2 = tails$, and $p(A_1) = p(A_2) = 0.5$. First we stipulate that:

- the probability of any realizable event is between zero and one: $p(A_i) \in [0 : 1]$ for all of the A_i in $i = 1, \ldots k$ in the sample space $\mathcal{S}$.

Next we require that one of the events must occur:

- some event happens with probability one: $p(\mathcal{S}) = 1$.

The next rule is a little more tricky and states that if we are considering whether any of a sub-collection of events can occur, say A_1 or A_3 or A_7, then the associated probability of this set is the sum of their individual probabilities. More formally:

- The probability of unions (collections) of n "pairwise disjoint" (that is, non-overlapping) events is the sum of their individual probabilities: $p(\bigcup_{i=1}^{n} A_i) = \sum_{i=1}^{n} p(A_i)$

(where $\bigcup$ means "union"). So $p(A_1 \bigcup A_3 \bigcup A_7) = p(A_1) + p(A_3) + p(A_7)$. To further illustrate this last rule return to flipping a fair coin from above: "the probability that we get heads *or* tails must be one, $p(A_1 \bigcup A_2) = p(heads) + p(tails) = 1$. In total these rules are called the *Kolmogorov probability axioms* (Kolmogorov, 1933), and they codify the modern definition of a probability function. Before this definition was institutionalized by mathematicians, authors would present their version of these statements at the beginning of the paper and prove or derive some objective based on their version, which meant that readers had to exert extra effort to understand the work than was actually necessary.

The next important principle to understand is *conditional probability*, which is simply the idea that prior information may change the probability of interest. For example if we are interested the probability of rolling a ⚅ with a far die, and we know in advance that the outcome is an even number, then sample space reduces from all six outcomes to the set {⚁, ⚃, ⚅} so now $p(⚅) = \frac{1}{3}$ instead of $\frac{1}{6}$. In more formal notation for event A, the unconditional (marginal) probability is $p(A)$ and the conditional probability given a prior event B is $p(A|B)$, "the probability of event A *given* event B has occurred." If B is meaningful in the calculation of $p(A)$ then $p(A) \neq p(A|B)$ and we would not want to ignore this information. This is calculated mathematically by:

$$p(A|B) = \frac{p(A \cap B)}{p(B)}, \tag{2.1}$$

given that $p(B) \neq 0$. Here $\cap$ in the numerator indicates "joint" or "intersection" meaning that the two events both occur, "A and B." Henceforth $\cap$ will be replaced with a comma, as in $p(A, B)$, which is more common in statistical notation for the joint probability of A and B. For the die rolling experiment we know that $p(⚅, \text{even}) = \frac{1}{6}$ because the only way *both* could occur is if we roll a ⚅. Since we know that the probability of rolling an even number is $\frac{1}{2}$, then the calculation of the conditional probability previously is calculated by:

$$p(A|B) = \frac{p(A, B)}{p(B)} = \frac{1/6}{1/2} = \frac{1}{3}.$$

Obviously this principle applies to more elaborate settings as we will see with Bayesian inference. The important idea to remember is that if we have prior information relevant to a probability calculation we would like to perform, then it will affect the resulting probability number when conditioned upon.

Conditional probability is *order dependent* meaning that the probability of event A given that event B has occurred is not the same as the probability of event B given that event A has occurred:

$$p(A|B) \neq p(B|A) \text{ and therefore from above } \frac{p(A, B)}{p(B)} \neq \frac{p(B, A)}{p(A)},$$

although

$$p(A, B) = p(B, A)$$

because order does not matter for joint probability statements just like "and" does not imply an order in English language statements. Given this fact we can write the two conditional probability statements accordingly and rearrange:

$$p(A|B) = \frac{p(A, B)}{p(B)} \qquad\qquad p(B|A) = \frac{p(A, B)}{p(A)}$$
$$p(A|B)p(B) = p(A, B) \qquad\qquad p(B|A)p(A) = p(A, B).$$

Therefore:

$$p(A|B)p(B) = p(B|A)p(A) \quad \longrightarrow \quad p(A|B) = \frac{p(A)}{p(B)}p(B|A), \tag{2.2}$$

meaning that we can switch the order of conditioning by multiplying with the ratio of marginal distributions in the order of the left-hand-side. This is so important that it has a special name, *Bayes' Law* from Bayes (1763), although we know that Laplace (1811) independently discovered this idea around the same time. In fact, this statement is the core of Bayesian inference in statistics as will be covered in detail in Section 4. The important idea to remember is that order is critical in conditional probability statements, and there is even a name for misunderstanding this principle called the "prosecutor's fallacy" since courtroom arguments sometimes make this exact mistake.

Consider the following example from the early months of the Covid pandemic as reported in *The New York Times* (August 6, 2020) for the city of New York that previous spring at the height of their pandemic emergency. The reported probabilities in the city population for having had Covid (*Cov*) and therefore having antibodies, along with test accuracy, are for randomly chosen resident:

- the estimated probability of Covid infection: $p(Cov) = 0.10$ (*prevalence*)
- the probability of correct positive (*Pos*) classification: $p(Pos|Cov) = 0.875$ (*sensitivity*)
- the probability of correct negative ($Pos^{\texttt{C}}$) classification: $p(Pos^{\texttt{C}}|Cov^{\texttt{C}}) = 0.975$ (*specificity*), where the "C" in the exponent denotes the complement of the associated probability, meaning $p(event^{\texttt{C}}) = 1 - p(event)$.
- Now suppose we want $p(Cov|Pos)$, from:

$$p(Cov|Pos) = \frac{p(Cov)}{p(Pos)}p(Pos|Cov),$$

  which is the probability of actual Covid infection given a positive test result.

To make this Bayes' Law calculation we need the marginal $p(Pos)$, which we do not immediately have. It can be obtained from the *Law of Total Probability*, which states that here there are only two ways to test positive (since this example has only two outcomes), testing positive with having Covid and testing positive not having had Covid. Using this fact, the definition of conditional probability, and complementation, we can calculate this marginal:

$$\begin{aligned} p(Pos) &= p(Pos, Cov) + p(Pos, Cov^{\texttt{C}}) \\ &\qquad \text{[from the Law of Total Probability]} \\ &= p(Pos|Cov)p(Cov) + p(Pos|Cov^{\texttt{C}})p(Cov^{\texttt{C}}) \\ &\qquad \text{[from turning joints into conditionals times marginals]} \\ &= p(Pos|Cov)p(Cov) + [1 - p(Pos^{\texttt{C}}|Cov^{\texttt{C}})]p(Cov^{\texttt{C}}) \\ &\qquad \text{[using complementation in the second term]} \\ &= (0.875)(0.10) + (1 - 0.975)(1 - 0.10) = 0.11. \end{aligned}$$

Now we have all of the ingredients for the Bayes' Law calculation:

$$p(Cov|Pos) = \frac{p(Cov)}{p(Pos)} p(Pos|Cov) = \frac{0.10}{0.11}(0.875) = 0.7954545.$$

This means that the probability that this individual has had Covid given a positive test classification is approximately 0.80. From the principle of complementation it also means that about 20% of those with a positive test classification never had the disease. These are not numbers that would make epidemiologists or policy makers particularly happy, but this was an extraordinary time and location for this disease.

Consider using computer simulation to ask probability questions. In traditional basic statistics courses it is common to see exercises that ask for probabilities over sub-regions of the support of a distribution. Most commonly this is tail values of a normal distribution (z-scores), but we can make it as general as we want. The key method introduced here, *Monte Carlo simulation* dating back to the earliest era of digital computing, is based on generating data according to a known distribution and manipulating the values arithmetically to compute desired quantities. Since computers can easily generate large numbers of specified random variables, the Law of Large Numbers and the Central Limit Theorem are applicable and we can replace analytical human calculations with easier computational work. Suppose we want to calculate the density of a normal distribution with mean 3 and standard deviation 2 that is above zero? This is $p_{normal}(y > 0|\mu = 3, \sigma = 2)$. A recipe is to generate one million (1M) values from this distribution, count the number of values above zero, and divide that by the number generated. This is three lines of `R` or `Python` code (a single line if one wants to be clever), and returns the value 0.933. This could be done more directly with specific functions in these languages, and this is an over-simplified use of Monte Carlo simulation. We provide this example here in the adjacent code boxes as a way to introduce simulation based calculation of probabilities with samples, which is a critical and necessary tool in more complex settings for Bayesian inference.

R Code for Simulated Probability Calculations

```
n.sims <- 1000000
y <- rnorm(n.sims,mean=3,sd=2)
length(y[y>0])/n.sims
```

Python Code for Simulated Probability Calculations

```
import numpy as np
n_sims = 1000000
```

```
y = np.random.normal(3,2,n_sims)
sum(y>0)/n_sims
```

The purpose of this section has been to introduce (or review) the necessary basic probability theory for continuing our studies in Bayesian inference. Probability is a deep area of study in mathematics and is pervasive in statistics, data science, and machine learning. Many readers will eventually want to continue studies in this area.

3 What Is a Likelihood Function and Why Care

Many statistics are produced by closed-form mathematical expressions that are well understood in both their calculation and their subsequent properties. The best known form of these is the calculation of the estimate of the vector of regression parameters for a multivariate linear model, which is simply $\hat{\beta} = (\mathbf{X}'\mathbf{X})^{-1}\mathbf{X}'\mathbf{y}$ where $\mathbf{X}$ is the matrix of explanatory variables down columns with a leading column of 1s and $\mathbf{y}$ is the vector of outcomes (bold indicates a vector or matrix structure). This is literally the most studied expression in statistics and we know everything about its properties. Another class of statistics are produced by assuming a *distributional property*, which leads to a functional form to be numerically maximized in order to produce estimates with known and optimal properties. The latter parametric approach is the topic of this section and a vast number of models in modern statistics need to use this approach to produce useful results for the researcher.

Random variables are defined by their *probability mass functions* for the discrete case (PMF), or their *probability density functions* for the continuous case (PDF). These distributions are probability functions, in the sense of the last section, that describe assumed variation in some random variable, say Y over a specific range of values (support), and are typically conditional on parameters, say θ. For the discrete case we specify $p(Y = y)$ for the probability that the random variable Y takes on some specific realization y. So in the case of flipping a fair coin we say $p(Y = 1) = 0.5$ and $p(Y = 0) = 0.5$, where 1 indicates heads and 0 indicates tails (this numerical assignment is standard but arbitrary). In the continuous case we consider $f(y)$ and often focus intervals on the real line ($\Re$) since exact numbers do not make sense as specific outcomes since they have probability zero of occurring on the real line. When the random variable Y is conditioned on parameters, that can be known or unknown, then we want to use conditional probability as discussed in the last section. Suppose that the random variable under study, Y, is distributed Exponential: $p(\mathrm{Y}|\theta) = \theta e^{-\theta \mathrm{Y}}$ (the "rate" version) with support $(0{:}\infty)$. This is a conditional probability statement since the distribution of Y is overtly dependent on the value of θ as in the definitional

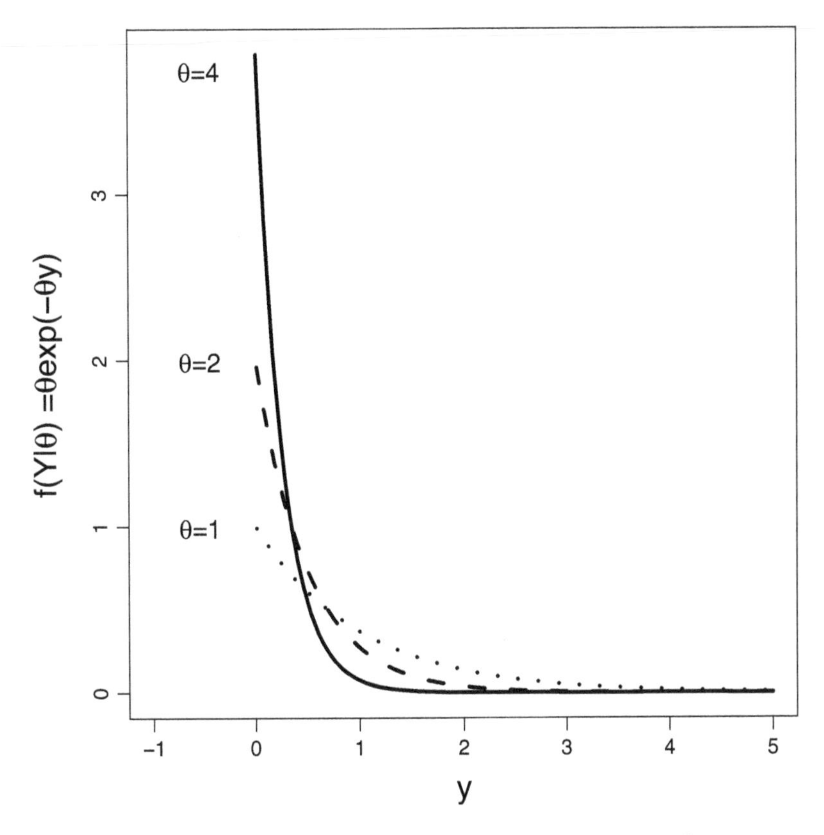

Figure 1 The exponential distribution for different θ values

form in (2.1). Note that this is a PDF since values of Y are defined over the real line from zero to positive infinity, and this makes the exponential distribution useful for modeling time to an event. Several versions with different values of θ are shown in Figure 1.

For the moment consider a "generic" conditional distribution for the random variable Y, $p(Y|\theta)$, but we observe an independent sample of n of them from the same data generating process conditioned on the same θ: $y_1, y_2, \ldots, y_n$. This is called an "independent, identically distributed" set of data, usually abbreviated IID or iid since they come from the same distribution but are produced without being conditional on each other in any way. Classic data that do not have this property come from a time series where typically serially observed values are conditional on previous observations, and conditions (conditional parameters) are likely to change over time like stock prices or agricultural output.

As a quick illustration, return to the coin flipping example but suppose we flipped the fair coin three times. The observed set of heads and tails is clearly an IID dataset, denoted here as $\mathbf{y} = (y_1, y_2, y_3)$ with partial PMF $p(Y = y), y = 0, 1$.

For just one of these events the probability of an observation is fully described by the PMF. Now ask the question: what is the probability of observing the data (H, T, H)? It is intuitively given by:

$$p(H, T, H) = p(Y = 1) \times p(Y = 0) \times p(Y = 1) = (0.5)(0.5)(0.5) = 0.125$$

(with standard encoding of heads and tails). We know that it is a fair coin so let us specify a conditional parameter to remind us of this fact: $p(Y = 1|\theta = 0.5) = 0.5$. Since only two things can happen we also know for a fact that $p(Y = 0|\theta = 0.5) = 0.5$, but this is redundant information and we only need the first statement to have full information. Returning to the generic description of the observed data, $\mathbf{y} = (y_1, y_2, y_3)$, we can more generally state that:

$$p(\mathbf{y}|\theta = 0.5) = p(y_1|\theta = 0.5)p(y_2|\theta = 0.5)p(y_3|\theta = 0.5) = \prod_{i=1}^{3} p(y_i|\theta = 0.5),$$

where the product notation ($\prod$) is a convenience to make such statements with arbitrary and possibly large sample sizes manageable notationally. Because it is a fair coin, $\theta = 0.5$, with only two sides, the probability of observing any specific n sized sample will always be equal to $(0.5)^n$, which is oversimplified for our intentions. A slightly more interesting variant would be a "biased" coin with, say, $p(Y = 1|\theta = 0.75)$. Now the probability of observing the sequence from above changes to:

$$p(H, T, H) = p(Y = 1) \times p(Y = 0) \times p(Y = 1) = (0.75)(0.25)(0.75) = 0.140625.$$

Note that this numeric value is higher than before. This is because it is more likely to draw a head and we did draw more heads than tails.

Now consider that we do not know the value of θ before sampling the data but we are willing to impose a parametric assumption with some form of $p(Y = y|\theta)$. For the exponentially distributed random variable earlier we specified that $p(\mathrm{Y}|\theta) = \theta e^{-\theta \mathrm{Y}}$. In most statistical modeling we make such a *parametric assumption* at the beginning of the analysis based on previous information, standard practice, or exploratory criteria (although there exist "nonparametric" modeling approaches, also called "semiparametric," in practice). Of course not knowing the true value of θ is not only more realistic, it adds substantially to the challenge of analyzing the data and calculating probabilities as we did previously. So now for an arbitrary set of n data values:

$$p(\mathbf{y}|\theta) = p(y_1|\theta)p(y_2|\theta) \cdots p(y_n|\theta) = \prod_{i=1}^{n} p(y_i|\theta). \tag{3.1}$$

This would be just as easy to mathematically manipulate as the earlier examples, even for large n, except that we do not know θ. A typical statistical problem is to observe the sample of y values and use them to estimate some unknown population parameters such as θ. In fact this is the exact definition of "inference" in statistics. For example, most basic statistics texts start with using a sample of y values to calculate $\bar{y}$ to estimate the true population mean, μ_y. Importantly, once the vector of y values is observed then it is *fixed* for our purposes. We can go get coffee for ten minutes, or go on vacation for two weeks, and when we return and look on our computer's hard drive it will still be the same (barring something catastrophic of course). So what is unknown (and possibly random) is θ. In two of the most important papers in the history of statistics Fisher (Fisher, 1922, 1925), considered this situation and decided to first change the notation of (3.1) to reflect that it is the $\mathbf{y}$ that is known and the θ that is target of our inquiry:

$$L(\theta|\mathbf{y}) = \prod_{i=1}^{n} p(y_i|\theta). \tag{3.2}$$

This is a notational sleight of hand since we have discussed conditional probability as going in the opposite direction and in the last section showed that order critically matters in conditional probability, but it is expressed this way to emphasize what is known and what is unknown. So now attention is focused on estimating the best value of the unknown θ given the observed and fixed vector $\mathbf{y}$. So what we want now is the value of θ, denoted $\hat{\theta}$ ("theta hat"), that is *most likely to have generated the observed data* given an assumed parametric form. For this reason the "L" in (3.2) stands for "Likelihood" and (3.2) is called a *likelihood function*. But it is important to remember that technically a likelihood function is just the product of the n PMFs or PDFs of the observed IID data.

Now our job is to find the best $\hat{\theta}$ for (3.2), that is, the value of y that is "most likely to have generated the observed data." This is called the *Maximum Likelihood Estimate* (MLE). For of the most common forms of $p(y|\theta)$ used in practice the form of $L(\theta|\mathbf{y})$ is concave to the x-axis as shown in Figure 2. This means that the function has a single unique maximum value of $L(\theta|\mathbf{y})$ for a given observed dataset. Here is where the big intellectual leap occurs. This unique value, $\hat{\theta}$, is *the single value that is most likely to have produced the observed data*, indicated by the dotted line in Figure 2. Therefore if the sample data is truly representative of the population data and the parametric assumption is correct, then this is the best estimate of the unknown population value of θ. In addition, this estimator has been proven to have optimal properties under common circumstances (Birnbaum, 1962). This is a very subtle and deep idea and most readers (at all levels!) do not fully grasp the theoretical importance right away.

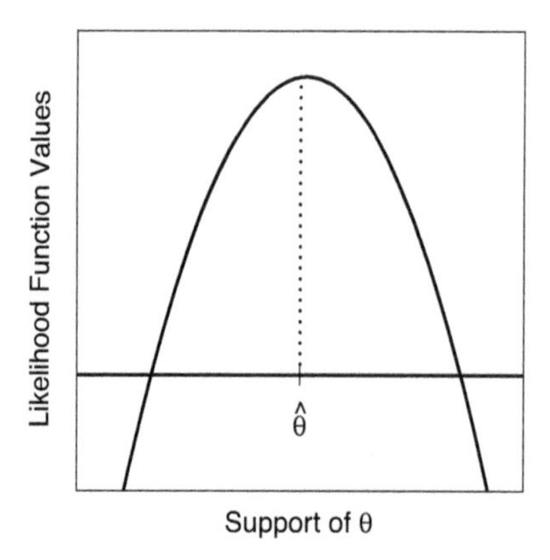

Figure 2 A generic likelihood function

Perhaps the best way to add intuition is to work through a detailed and realistic example. Consider a standard model for evaluating data that are counts: nonnegative integers without an upper bound. Another way of thinking of such count data is in terms of durations: the time waiting for some event of interest. If the probability of an event is *proportional to the length of the wait*, then the number of events in a given time period can be modeled with the Poisson distribution:

$$p(y|\theta) = \frac{e^{-\theta}\theta^{y}}{y!}, \quad y \in \mathcal{I}^{+},\ \theta \in \Re^{+}. \tag{3.3}$$

which is read as the "the probability that exactly y events occur given parameter θ." Notationally $y \in \mathcal{I}^{+}$ indicates that the support of the outcome variable y is over the nonnegative integers, and $\theta \in \Re^{+}$ indicates that the support of θ is over the positive real line, $\Re^{+}$. The assumption of proportionality is usually quite reasonable because over longer periods of time the event has more "opportunities" to occur. Here θ is called the *intensity parameter* and gives the mean (expected) number of events and the (expected) variance. The concept of expectation is described in detail in Section 6. To use the Poisson probability model for counts, we need to assume:

1. **Non-Simultaneity:** Two events cannot occur at *exactly* the same time.
2. **IID:** Events in different time segments are independent and identically distributed.
3. **Proportionality:** For small time periods, the probability of an event is proportional to the length of time passed in the period so far, and is not dependent on the number of previous events in this period.

This is a basic model for counts and there are many variants when the aforementioned assumptions are violated and for other complexities.

Now assume that we have a vector of counts, $\mathbf{y}$, and we want the MLE of the unknown θ parameter. The likelihood function is created from the joint distribution of the observed data as shown earlier:

$$L(\theta|\mathbf{y}) = \prod_{i=1}^{n} \frac{e^{-\theta}\theta^{y_i}}{y_i!} = \frac{e^{-\theta}\theta^{y_1}}{y_1!}\frac{e^{-\theta}\theta^{y_2}}{y_2!}\cdots\frac{e^{-\theta}\theta^{y_n}}{y_n!} = e^{-n\theta}\theta^{\sum y_i}\left(\prod_{i=1}^{n} y_i!\right)^{-1}.$$

Suppose now that we have the count data: $\mathbf{y} = (5, 1, 1, 1, 0, 0, 3, 2, 3, 4)$, then the likelihood function from plugging in these values previously is:

$$L(\theta|\mathbf{y}) = \frac{e^{-10\theta}\theta^{20}}{207360},$$

which is the probability of observing *this* exact sample. With likelihood function calculations is often easier to deal the logarithm:

$$\begin{aligned}\log L(\theta|\mathbf{y}) = \ell(\theta|\mathbf{y}) &= \log\left(e^{-n\theta}\theta^{\sum y_i}\left(\prod_{i=1}^{n} y_i!\right)^{-1}\right)\\ &= -n\theta + \sum_{i=1}^{n} y_i \log(\theta) - \log\left(\prod_{i=1}^{n} y_i!\right) \qquad (3.4)\end{aligned}$$

where the "$\log L$" and "ℓ" notation are both commonly used. For our small example this is numerically:

$$\ell(\theta|\mathbf{y}) = -10\theta + 20\log(\theta) - \underbrace{\log(207360)}_{12.242}. \qquad (3.5)$$

Importantly, for the family of functions that we will use the likelihood function and the log-likelihood function have the same mode (maximum of the function) for θ, and they are both guaranteed to be concave to the x-axis. This is illustrated for this example in Figure 3. The property of having only one mode and no minima is important because our calculus tool (taking the derivative, setting equal to zero) finds points with a minima or maxima without regard for which, so in this case we know what we are finding.

So now let's use freshman calculus to find out the maximum of the log likelihood function. This is at the θ point when first derivative of $\ell(\theta|\mathbf{y})$ equals zero, where the dotted line in Figure 2 touches the curve. This means that we take the first derivative with regard to the parameter of interest denoted $\frac{d}{d\theta}$, set the resulting equation equal to zero, and solve for θ. Here the statement $\frac{d}{d\theta}\ell(\theta|\mathbf{y}) \equiv 0$ is unsurprisingly called the *Likelihood Equation*. For the numerical example we start with $\ell(\theta|\mathbf{y})$ and take the derivative (which uses the

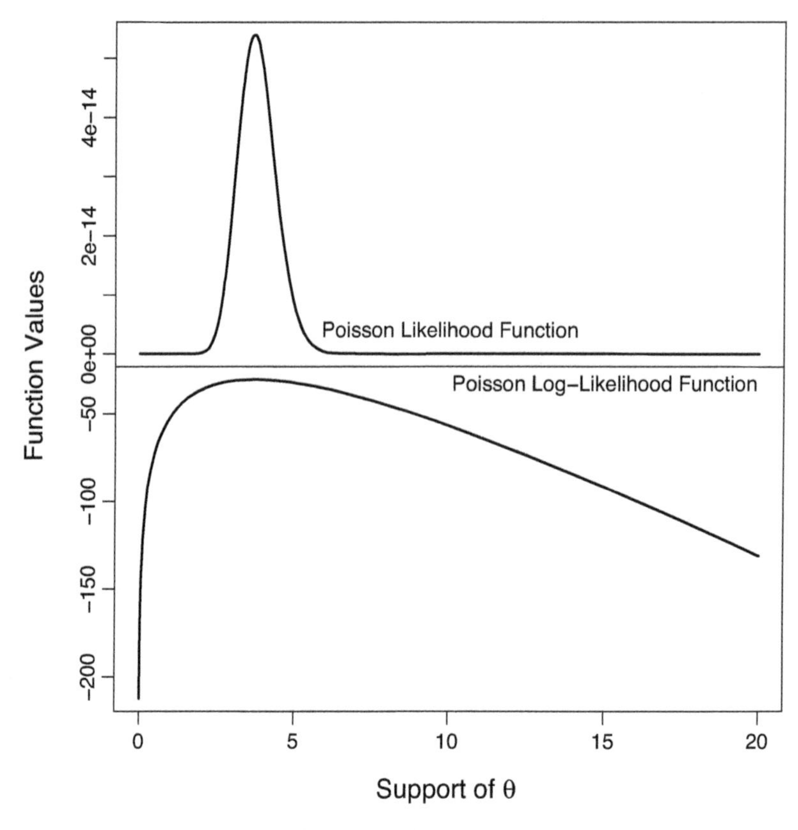

Figure 3 Poisson likelihood function and log-likelihood function

exponent rule in calculus: $\frac{d}{dx}(\text{some constant})x^k = (\text{some constant})kx^{k-1}$), and setting equal to zero gives:

$$\frac{d}{d\theta}\ell(\theta|\mathbf{y}) = \frac{d}{d\theta}(-10\theta + 20\log(\theta) - 12.242) = -10 + 20\theta^{-1} \equiv 0, \qquad (3.6)$$

where we use the additional rules that $\frac{d}{dx}(\text{some constant alone}) = 0$, and $\frac{d}{dx}\log(x) = x^{-1}$. So that $20\theta^{-1} = 10$, and therefore the MLE is $\hat{\theta} = 2$ (note the hat indicating that this is now an estimate). This is the most likely θ value with the Poisson PMF to have generated the data $\mathbf{y} = (5, 1, 1, 1, 0, 0, 3, 2, 3, 4)$. Note that the mathematical expression $\equiv 0$ means "set equal to zero," as opposed to $= 0$ which means "is equal to zero."

In the more general sense with arbitrary data we can perform the same process to get a generic form of the Poisson MLE:

$$\ell(\theta|\mathbf{y}) = -n\theta + \sum_{i=1}^{n} y_i \log(\theta) - \log\left(\prod_{i=1}^{n} y_i!\right)$$

[take the first derivative with respect to θ]

$$\frac{d}{d\theta}\ell(\theta|\mathbf{y}) = -n + \frac{1}{\theta}\sum_{i=1}^{n} y_i - 0 \equiv 0$$

[algebraically rearrange]

$$\hat{\theta} = \frac{1}{n}\sum_{i=1}^{n} y_i = \bar{\mathbf{y}}$$

Note that it is *not* true that the MLE is always the data mean; this is just a special property of the Poisson PMF. So to summarize the general process, perform the following steps:

1. identify the PMF or PDF applicable to the outcome
2. create the likelihood function from the joint distribution of the observed data
3. change to the log-likelihood for convenience
4. take the first derivative with respect to the parameter of interest
5. set this equation equal to zero
6. solve algebraically for the MLE.

Of course with more complicated forms this process is done in software such as with generalized linear models regression (logit, probit, gamma, etc.) through a process called Iteratively Weighted Least Squares; see Gill and Torres (2019).

R Code to for the Poisson MLE Example

```
# A POISSON LIKELIHOOD AND LOG-LIKELIHOOD FUNCTION
llhfunc<-function(X,p,do.log=TRUE) {
    d <- rep(X,length(p))
    q.vec <- rep(length(X),length(p))
    p.vec <- rep(p,q.vec)
    d.mat <- matrix(dpois(d,p.vec,log=do.log),
        ncol=length(p))
    if (do.log==TRUE) apply(d.mat,2,sum)
    else apply(d.mat,2,prod)
}

y.vals<-c(1,3,1,5,2,6,8,11,0,0)

# EXAMPLE RUN FOR TWO POSSIBLE VALUES OF THETA: 4 AND 30
llhfunc(y.vals,c(4,30))

# USE THE R CORE FUNCTION FOR OPTIMIZING,
```

```
# par=STARTING VALUES,
# control=list(fnscale=-1) INDICATES A MAXIMIZATION,
# bfgs=QUASI-NEWTON ALGORITHM
mle <- optim(par=1,fn=llhfunc,X=y.vals,
    control=list(fnscale=-1),method="BFGS")

# MAKE A PRETTY GRAPH OF THE LOG AND NON-LOG VERSIONS
ruler <- seq(from=.01, to=20, by= .01)
poisson.ll <- llhfunc(y.vals,ruler)
poisson.l <- llhfunc(y.vals,ruler,do.log=FALSE)

par(oma=c(3,3,1,1),mar=c(0,0,0,0),mfrow=c(2,1))
plot(ruler,poison.l,type="l",xaxt="n",lwd=3)
text(mean(ruler),mean(poison.l),
    "Poisson Likelihood Function")
plot(ruler,poison.ll,,type="l",lwd=3)
text(mean(ruler)+5,mean(poison.ll)/2,
    "Poisson Log-Likelihood Function")
```

PYTHON Code to for the Poisson MLE Example

```
import numpy as np
from scipy.stats import poisson
from scipy.optimize import minimize
import matplotlib.pyplot as plt

# POISSON LIKELIHOOD AND LOG_LIKELIHOOD FUNCTION
def llhfunc(X, p, do_log=True):
    lX, lp = len(X), len(p)
    d = np.tile(X, lp)
    u = np.repeat(p, lX)
    p_pmf = [poisson.pmf(d[i:i+lX], u[i:i+lX])
             for i in range(0, lX*lp, lX)]
    d_mat = np.log(p_pmf).T if do_log else \
            np.array(p_pmf).T
    return np.sum(d_mat, axis=0) if do_log else \
           np.prod(d_mat, axis=0)

y_vals = np.array([1, 3, 1, 5, 2, 6, 8, 11, 0, 0])
```

```
# EXAMPLE RUN FOR TWO POSSIBLE VALUES OF THETA: 4 AND 30
llhfunc(y_vals,np.array([4, 30]))

# USE minimize() FROM scipy.optimize
mle = minimize(fun=lambda p: -llhfunc(y_vals, p),
               x0=1,
               method="BFGS")
# fun=lambda p: -llhfunc(y_vals, p): DEFINE FUNCTION
# x0=1: STARTING VALUES
# method="BFGS": QUASI-NEWTON ALGORITHM

ruler = np.arange(.01, 20.01, .01)
poison_ll = llhfunc(y_vals, ruler)
poison_l = llhfunc(y_vals, ruler, do_log=False)

fig, axs = plt.subplots(2, 1, figsize=(8, 12))
axs[0].plot(ruler, poison_l, linewidth=3)
axs[0].annotate('Poisson Likelihood Function',
                xy=(np.mean(ruler),
                    np.mean(poison_l)),
                xytext=(np.mean(ruler)-2,
                        np.mean(poison_l)))
axs[0].tick_params(which='both',
                   bottom=False,
                   labelbottom=False)
axs[1].plot(ruler, poison_ll, linewidth=3)
axs[1].set_xlabel('Support of $\Theta$')
axs[1].annotate('Poisson Log-Likelihood Function',
                xy=(np.mean(ruler),
                    np.mean(poison_ll)),
                xytext=(np.mean(ruler)+2,
                        np.mean(poison_ll)+20))
plt.subplots_adjust(hspace=0)
plt.show()
```

We can also measure the uncertainty of the MLE around this point. Recall that all reported statistics should be accompanied by an associated measure of uncertainty, usually a standard error. To obtain this measurement we will return to basic derivative calculus. When working with functions the first derivative measures slope of the tangent line at given points (zero was of interest in our case before) and the second derivative measures "curvature" of the function at a given point. The second derivative is obtained by taking a function's first derivative and performing the same operation again on that: taking $\frac{d}{d\theta}\ell(\theta|\mathbf{y})$ and repeating the differentiation process, $\frac{d}{d\theta}(\frac{d}{d\theta}\ell(\theta|\mathbf{y}))$. Getting this curvature is important because the more peaked the log-likelihood function is at the MLE, the more "certain" the data are about this estimator. Conversely the more diffuse the log-likelihood function around the MLE, the less the data are saying about the this estimate.

Going from curvature to a standard error of the MLE is relatively simple. The *square root of the negative inverse of the expected value of the second derivative is the standard error of the MLE.* We will save the details of expected value for Section 6, but the operation is very simple here. Calculate the first derivative:

$$\frac{d}{d\theta}\ell(\theta|\mathbf{y}) = -n + \frac{1}{\theta}\sum_{i=1}^{n} x_i$$

and then take another (now second) derivative of this form:

$$\frac{d^2}{d\theta^2}\ell(\theta|\mathbf{y}) = \frac{d}{d\theta}\left(\frac{d}{d\theta}\ell(\theta|\mathbf{y})\right) = -\theta^{-2}\sum_{i=1}^{n} x_i$$

again using the exponential rule in calculus. The expected value (estimate) of θ is the MLE (data mean for the Poisson case) allowing us to replace $\hat{\theta}$ with $\bar{\mathbf{y}}$, so:

$$\text{Var}(\hat{\theta}) = \left(\hat{\theta}^{-2}\sum_{i=1}^{n} x_i\right)^{-1} = \frac{\hat{\theta}^2}{\sum_{i=1}^{n} x_i} = \frac{\bar{\mathbf{y}}^2}{n\bar{\mathbf{y}}} = \frac{\bar{\mathbf{y}}}{n}.$$

Now we have the MLE for the general Poisson case, $\bar{\mathbf{y}}$, and its associated standard error, $(\mathbf{y}/n)^{-1/2}$. In our numerical example the standard error is then simply $\sqrt{2/10} = 0.4472$.

4 The Core of Bayesian Inference: Prior Times Likelihood

So far everything in this exposition has been leading up to this section, which is easily the most important one. How are Bayesian models constructed, and how is Bayesian inference performed? Here we will look at the way prior information and data are combined to provide inferences that are a balance between the two.

We start with broad high-level language giving the philosophical principles of Bayesian inference before proceeding to the mechanics and details of the process. The first, and perhaps most important, principle is that all unknowns are treated probabilistically, meaning they are assigned a conditional distributional statement $p(\cdot|\cdot)$, or a probability value $[0:1]$. This includes unknown parameters, missing data values, and even model choices. Thus *uncertainty* is always a *probabilistic* phenomenon. Specifically data enters the inference using probability models in the form of likelihood functions, which, recall are just the joint probability of the observed data using an assumed parametric form. Then all parameters (variables of interest or requirement) are assigned distributions with all relevant information prior to considering the observed data at hand: PMFs or PDFs depending on the level of measurement.

The Bayesian inferential process proceeds by using prior information combined data information to produce an updated distribution for each parameter of interest with the principle of inverse probability through conditional distributions presented in Section 2 as Bayes' Law. This provides a full probabilistic description that can be evaluated in many ways. The overall process can be then described in three steps:

1. Specify a probability model for unknown parameter values that includes some prior knowledge about the parameters if available.
2. Update knowledge about the unknown parameters by conditioning this probability model on observed data.
3. Evaluate the fit of the model to the data and the sensitivity of the conclusions to the assumptions.

We will, of course, go over these steps in detail in the process of this monograph. For now readers should have an appreciation for the considerable difference between Bayesian inference and the maximum likelihood process for estimation described in Section 3, even though both use a likelihood function.

4.1 The Core Bayesian Process

Bayesian inference is about updating. The core operation is taking a distribution that represents the current state of knowledge about some phenomenon of interest and updating it with recently acquired information to produce a new distribution that is more informed. Our starting point is (2.2) in Section 2, which stated $p(A|B) = \frac{p(A)}{p(B)}p(B|A)$, and can be thought of as updating A with new information B from a version of A with no conditioning, $P(A)$. Now replace this A with θ as some model parameter that we would like to estimate, and replace B with a more direct expression of the data, $\mathbf{y}$. Therefore $p(B|A)$ is a joint distribution of the data conditional on the parameter, the likelihood function: $L(\theta|\mathbf{y})$. Now Bayes' Law can be expressed in the real quantities of interest:

$$\pi(\theta|\mathbf{y}) = \frac{p(\theta)L(\theta|\mathbf{y})}{p(\mathbf{y})}, \tag{4.1}$$

where the $\pi()$ notation is just a reminder that the left-hand-side expression is different than the $p(\theta)$ on the right-hand-side. So far this is very intuitive, but how do we interpret $p(\mathbf{y})$? This is the unconditional probability of observing the data that we have in hand right now. Remember to a Bayesian once the data are observed they are fixed forever and no longer have a random (stochastic) characteristic, as opposed to a frequentist who has an unending stream of random data. So in a Bayesian sense then $p(\mathbf{y}) = 1$. So why is it even there if it is just an unimportant constant? One answer is that it is a constant that ensures that $\pi(\theta|\mathbf{y})$ sums or integrates to one so that it is a proper probability statement in the sense of the Kolmogorov probability axioms discussed in Section 2. So specifically, what would happen if we left it out is that $\pi(\theta|\mathbf{y})$ would not sum (PMF) or integrate (PDF) to 1 as a proper probability function should. If this denominator is actually not the number 1, then what is it? It is whatever number that in divisor makes the right-hand-side sum or integrate to one. It is easier for now to consider only the continuous case where all of the functions in (4.1) above are PDFs. If we want $p(\theta)L(\theta|\mathbf{y})$ to integrate to 1 then we need to divide it by the area it occupies under the curve over the support of θ on the x-axis labeled Θ as shown hypothetically in Figure 4. Remember $\mathbf{y}$ is fixed here so only θ has a distributional property. This area is calculated by the integral:

$$\int_{\Theta} p(\theta)L(\theta|\mathbf{y})d\theta = f(\mathbf{y}), \tag{4.2}$$

where: $\int$ is a curvy version of $\sum$ for an interval measured variable, Θ is the support of the variable θ on the x-axis, and $d\theta$ is a reminder that this calculation is done with respect to the variable θ ("the variable of integration"). Some of this notation feels like redundant information but it is a hint that integrals can be much more complex and nuanced with many variables and different types of support (Section 6). So if $p(\theta)L(\theta|\mathbf{y})$ integrated to 1.4, say, we would just divide it by 1.4 to make $\pi(\theta|\mathbf{y})$ a proper probability statement. So actually performing the integral (typically done computationally, if at all) would produce a number that we would use accordingly.

All of this discussion of the denominator leads to two forms of the key Bayesian inferential expression. First there is:

$$\pi(\theta|\mathbf{y}) = \frac{p(\theta)L(\theta|\mathbf{y})}{\int_{\Theta} p(\theta)L(\theta|\mathbf{y})d\theta}, \tag{4.3}$$

which is the primary definitional form that says that the posterior distribution is obtained with Bayes' Law from the prior distribution times the likelihood function with a scaling factor in the denominator. So we have a balance or

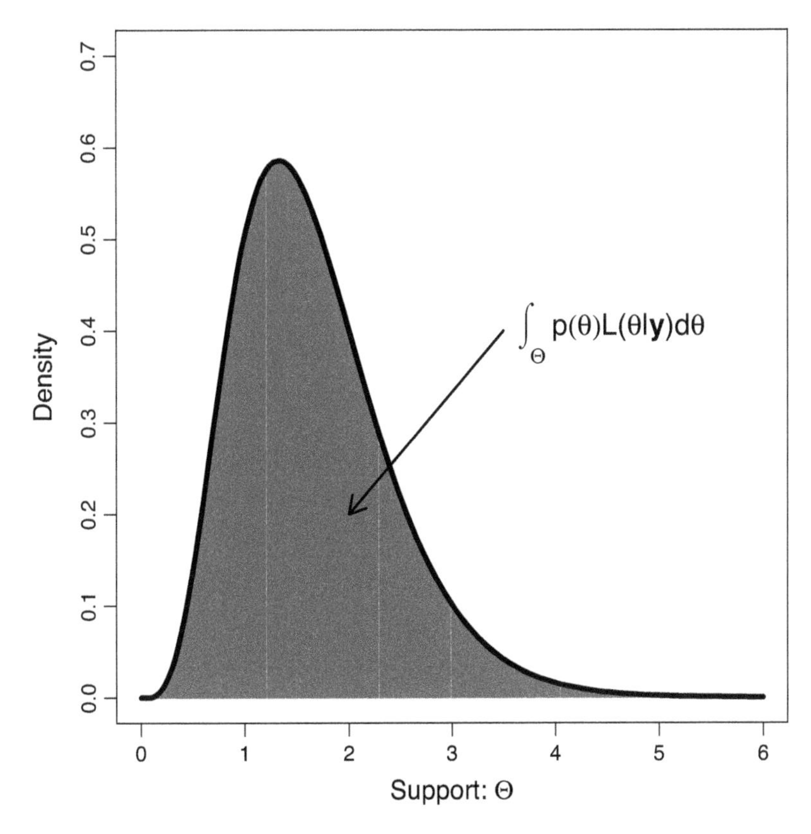

Figure 4 Integrated likelihood

compromise of prior information and data information that become posterior information. Now that we know that the denominator's main purpose is simply to make the posterior distribution a proper one and that this purely numeric information can be recovered at any time, it is convenient to drop it and use a proportional statement ($\propto$) for the relationship:

$$\pi(\theta|\mathbf{y}) \propto p(\theta)L(\theta|\mathbf{y}), \tag{4.4}$$

which is read as:

Posterior Probability $\propto$ Prior Probability $\times$ Likelihood Function.

This is the form that you see most often because it contains the core of the relationship and simply omits an inconvenience that can be recovered later after key calculations. The omitted denominator, called the "integrated likelihood," the "marginal likelihood," the "marginal probability of the data," or the "predictive probability of the data," seems unimportant here but it is actually very useful in other contexts, including the model comparison tool Bayes Factor.

From all of this we see the key fundamental principle of Bayesian inference: start with a prior distribution, $p(\theta)$, that is unconditional on the data at hand and reflects prior belief in the distribution stipulated, multiply it with the likelihood function, $L(\theta|\mathbf{y})$, where the data enters the equation, to produce the posterior distribution, $\pi(\theta|\mathbf{y})$, which is a distribution that reflects our latest knowledge about the phenomenon of interest. Notice that the result is a *distribution* not a point estimate, meaning that uncertainty is built-in to the results as shown by how peaked or flattened this posterior distribution appears. One powerful feature of this process is that in the future if additional data ($\mathbf{y}_2$) are observed then we can treat the current posterior based on the earlier data ($\mathbf{y}_1$) as a prior distribution in (4.4), and use the new data in a new likelihood function, and create an updated posterior to reflect the latest knowledge about θ:

$$\pi(\theta|\mathbf{y}_2) \propto \pi(\theta|\mathbf{y}_1)L(\theta|\mathbf{y}_2). \tag{4.5}$$

Plus, the final posterior created in this process is exactly the same as if we had both datasets ($\mathbf{y}_1, \mathbf{y}_2$) at the same time. This updating process can also be repeated as many times as data arrival permits with the same principle applying.

4.2 Mathematical Example of Posterior Calculation

To show how this process works with a specific model and actual social science data, return to the Poisson model from Section 2 where we developed the likelihood function for count data using a Poisson specification to provide inference for an unknown intensity parameter θ. Recall that the derived likelihood function for this setup with an n-length sample $\mathbf{y}$ was:

$$L(\theta|\mathbf{y}) = \exp[-n\theta]\theta^{\sum y_i}\left(\prod_{i=1}^{n} y_i!\right)^{-1}. \tag{4.6}$$

A convenient and flexible prior distribution for θ with the right support is the gamma distribution:

$$p(\theta|\alpha,\beta) = \beta^{\alpha}\Gamma(\alpha)^{-1}\theta^{\alpha-1}\exp[-\beta\theta],\ \alpha,\beta,\theta > 0 \tag{4.7}$$

given a shape parameter α, and rate parameter β (the latter alternately expressed as a scale parameter $1/\beta$). Also, the Gamma function ($\Gamma()$ earlier is the generalization of the factorial function ($k! = k \times (k-1) \times (k-2) \cdots 2 \times 1$) that can be applied to noninteger values($\Gamma(k) = \int_0^{\infty} t^{k-1}e^{-t}dt,\ k > 0$). The exponential distribution shown in Figure 1 is actually a special case of the gamma distribution where the α parameter is set to one. As discussed in Section 4, it is the researcher's responsibility to set the two "hyperparameter" values (α

and β) according to theory, convention, or desired vagueness. The posterior distribution for θ is given by the calculation starting from (4.4):

$$\begin{aligned}
\pi(\theta|\mathbf{y}) &\propto p(\theta)L(\theta|\mathbf{y}) \\
&\qquad \text{[plug (4.7) and (4.6) in that order]} \\
&= \beta^{\alpha}\Gamma(\alpha)^{-1}\theta^{\alpha-1}\exp[-\beta\theta] \times \exp[-n\theta]\theta^{\sum y_i}\left(\prod_{i=1}^{n} y_i!\right)^{-1} \\
&\qquad \text{[strip off all constants using proportionality]} \\
&\propto \theta^{\alpha-1+\sum_{i=1}^{n} y_i}\exp[-(\beta+n)\theta]. \qquad (4.8)
\end{aligned}$$

The second step is justified because we are already using proportionality by ignoring the denominator of Bayes' Law, and therefore these constants add no important information. They will be recovered in a later step to make posterior statements "proper" in the sense that $\pi(\theta|\mathbf{y})$ integrates to 1 as probability function should. From the second step we can also see that ignoring the constants for the moment makes the form of the posterior more clear and less cluttered. Notationally these Bayesian expressions are often given without all of the fixed parameters on the side of the conditional if the dependency is obvious as it is with (4.8). This expression can be more completely described as $p(\theta|\mathbf{y},\alpha,\beta)$ on the left side of $\propto$, but the conditionality is clear without this. In more complex specifications listing all of these hyperparameter and fixed parameter values often muddies the notation.

To further improve our intuition we can define:

$$\alpha^{\dagger} = \alpha + \sum_{i=1}^{n} y_i \qquad\qquad \beta^{\dagger} = \beta + n,$$

which means that we can express the posterior distribution in the last line of (4.8) as:

$$\pi(\theta|\mathbf{y}) \propto \theta^{\alpha^{\dagger}}\exp[-(\beta^{\dagger})\theta],$$

which is the kernel (the part of a distribution absent constants) of a gamma distribution different than the gamma prior given by (4.7), which means that if use we proportionality in the other direction (adding back the constants) we get:

$$\pi(\theta|\mathbf{y}) = (\beta^{\dagger})^{\alpha^{\dagger}}\Gamma(\alpha^{\dagger})^{-1}\theta^{\alpha^{\dagger}-1}\exp[-\beta^{\dagger}\theta]. \qquad (4.9)$$

This means that the posterior distribution of θ is a gamma form with parameters $\alpha^{\dagger}$ and $\beta^{\dagger}$ previously, and that we now know *everything* about it. So we can report its mean, median, variance, quantile variances, and more. For instance the mean and variance of gamma distributed random variable X are:

- $E[X] = \frac{\alpha}{\beta}$, rate version.
- $\text{Var}[X] = \frac{\alpha}{\beta^2}$, rate version.

The first expression earlier for the expected value will be derived in Section 6 as an example of an expected value calculation. It is also now a trivial exercise to plot it in R or Python. This is a particularly elegant result partly because it is a gamma to gamma conjugate relationship through the Poisson likelihood function. Conjugacy is a property of a set of specific Bayesian models where pairing the distributional form of the prior distribution with a specific likelihood function means that the posterior distribution has the same form as the prior, with different parameterization of course.

We can also think about the expected value (mean) of this posterior gamma distribution for the parameter θ in an informed way with some modest algebra using the aforementioned definition:

$$E[\theta|\mathbf{y}] = \overline{\theta|\mathbf{y}} = \frac{\alpha^\dagger}{\beta^\dagger} = \frac{\alpha + \sum_{i=1}^{n} y_i}{\beta + n} = \left[\frac{n}{\beta + n}\right]\bar{\mathbf{y}} + \left[\frac{\beta}{\beta + n}\right]\left(\frac{\alpha}{\beta}\right). \tag{4.10}$$

The concept of expected value will be discussed at length in Section 6, but for now think of it as simply the mean of the distribution of the random variable θ given the data $\mathbf{y}$. This rewrite of the posterior mean into a weighed sum of the data mean and the prior mean shows some important principles here and is present in all Bayesian inference. Consider earlier what happens as n gets very large moving towards infinity. The second term in the sum goes away because n is only in the denominator of the bracketed component (the second weight), and in the first term the bracketed component (the first weight) goes to one for any reasonable choice of the number β. So asymptotically (in the limit as the data size gets bigger) the posterior mean for θ converges to the data mean, $\bar{\mathbf{y}}$, contained in the first term, and the prior mean, α/β, in the second term becomes irrelevant. So the data wins over the prior in the limit, and this is true in every Bayesian specification no matter how complex. It also makes intuitive sense: if we have massive amounts of (presumably high quality) data then prior information should matter less or not at all. Conversely, when the data size is limited to a modest number we want to be able to rely on high quality prior information. Furthermore, this is exactly how science is supposed to work: a large quantity of reliable new information should change our knowledge about some phenomenon of interest relative to prior information, and modest amounts of new information may not do so.

Table 1 Active shooter incidents in the United States

Year	2000	2001	2002	2003	2004	2005	2006	2007
Count	3	10	7	12	5	11	12	14
Year	2008	2009	2010	2011	2012	2013	2014	2015
Count	9	19	27	13	20	18	19	19
Year	2016	2017	2018	2019	2020	2021		
Count	19	31	30	30	40	61		

Note: Counts by Year in the United States.

Source: Pew Research Center, April 6, 2023, from FBI statistics.

4.3 Empirical Example of Posterior Calculation

Consider describing active shooter incidents by year in the United States from 2000 to 2021. The report cautiously notes that the definition of "mass shooter incident" is not fully agreed upon and therefore uses instead the FBI definition of active shooter incidents: "one or more individuals actively engaged in the killing or attempting to kill people in a populated area" to produce the count data reproduced in Table 1 with $n = 22$, and $\sum y_i = 429$. It turns out that now there is very little calculation to be done since we have a full recipe for the posterior of interest from (4.9) with the parameter definitions in (4.2).

Our first job is to determine a reasonable prior form for θ, which means selecting the two parameters in (4.7). A reasonably diffuse and conservative choice is $\alpha = 14$ and $\beta = 2$ for a prior mean and variance of 7 and 3.5, respectively, reflecting a much lower level of active shooter incidents before the year 2000. So stipulating this prior is an implicit Bayesian hypothesis test that these incidents have increased in the last two decades over previous eras. This prior decision leads to the simple conjugate update described in (4.9) on 25 to produce the gamma distribution posterior for θ:

$$\mathcal{G}_{\text{prior}}(14,2) \longrightarrow \mathcal{G}_{\text{posterior}}(14+429, 2+22) = \mathcal{G}(443,24),$$

meaning that the updated posterior mean and variance of θ are 18.46 and 0.77 respectively. This progression from prior through data to posterior is shown in Figure 5. The effect of the data on the prior is not only to move the center of the distribution but also to decrease the variability indicating that the data are informative in addition to what is contained in the prior. We can also compare quantiles using the `R` function `qgamma` or the `Python` function `gdtr`, as shown in Table 2.

Table 2 Quantiles from the prior and posterior, active shooter model

Quantile	0.05	0.25	0.50	0.75	0.95
Prior Distribution	4.232	5.664	6.834	8.155	10.334
Posterior Distribution	17.040	17.859	18.444	19.042	19.924

R Code for the Gamma Poisson Update

```
y <- c(3,10,7,12,5,11,12,14,9,19,27,13,20,18,19,19,19,
       31, 30,30,40,61)
par(mar=c(6,6,2,2), cex.lab=1.5)
ruler <- seq(from=0,to=25,length=300)
alpha <-14; beta <- 2
plot(ruler,dgamma(ruler,alpha,beta),type="l",
    ylim=c(0,0.5),lwd=3, col="grey70",
    ylab="Posterior Density",xlab="Support")
lines(ruler,dgamma(ruler,alpha+sum(y),beta+length(y)),
    lwd=4)
text(7,0.25,"Prior",col="grey70",cex=1.5,adj=0.5)
text(18.46,0.485,"Posterior",col="grey10",cex=1.5,
    adj=0.5)
```

Python Code for the Gamma Poisson Update

```
import matplotlib.pyplot as plt
import numpy as np
from scipy.stats import gamma

y = np.array([3,10,7,12,5,11,12,14,9,19,27,13,
              20,18,19,19,19,31,30,30,40,61])
ruler = np.linspace(0, 25, num=300)
alpha = 14; beta = 2

# GAMMA DISTRIBUTION IN scipy.stats USES SCALE FOR
# THE SECOND PARAMETER: 1/rate
d_val_prior = gamma.pdf(ruler, alpha, scale=1/beta)
d_val_post = gamma.pdf(ruler, alpha+sum(y),
                       scale=1/(beta+len(y)))
```

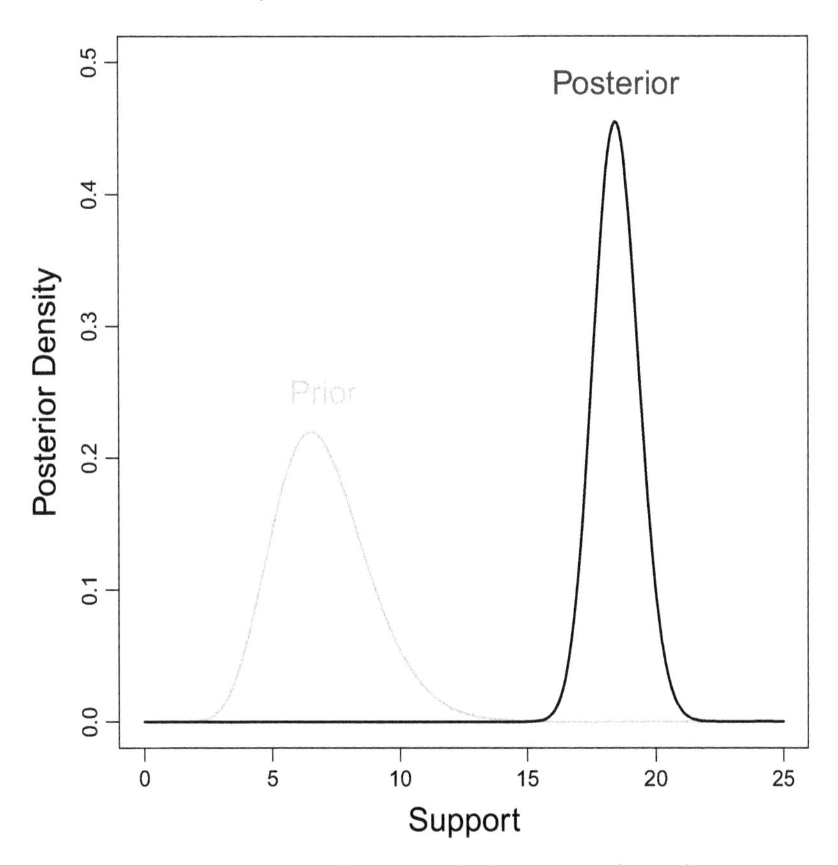

Figure 5 Gamma Poisson model prior to posterior update

```
fig, ax = plt.subplots(figsize=(8, 6))
ax.plot(ruler, d_val_prior, color='grey',
        linewidth=3, label="Prior")
ax.plot(ruler, d_val_post, color='black',
        linewidth=4, label="Posterior")
ax.set_xlabel('Support')
ax.set_ylabel('Density')
ax.text(6.5, 0.235, 'Prior', color='grey',
        fontsize=12, ha='center')
ax.text(18.46, 0.475, 'Posterior', color='black',
        fontsize=12, ha='center')
plt.ylim([0, 0.51])
plt.show()
```

5 Prior Probabilities and the Progression of Human Knowledge

Prior to posterior inference is how scientific knowledge accumulates, and prior information abounds in the social sciences. We have hundreds of years of self-aware study of the human condition and human interactions, both qualitative and quantitatively. In fact it seems daft to ignore deep literature, common wisdom, recorded history, and researcher intuition when creating statistical models to understand social science phenomena. Prior specifications can come from previous studies, conflicting theories in a given literature, researcher experience, nonparametrics and other data-oriented sources, diagnostic objectives, expert judgments, and even previous posterior distributions.

As discussed, the Bayesian process of inference starts with assigning prior distributions for unknown parameters. *Prior distributions are necessary in Bayesian models.* These unknown parameter distributions are operationalized with observed explanatory variables in a simple model. Such prior distributions range from very informative descriptions based on previous research in the field to purposefully vague and uncertain forms that reflect high levels uncertainty or possibly previous ignorance. It is important to notice that the prior distribution is not an inconvenience imposed on the researcher by the treatment of unknown quantities. It is instead an opportunity to include existing knowledge systematically in the model. Such prior information can include: qualitative, narrative, statistical, and intuitive information.

The use of prior distributions has been controversial at various times in the history of statistics. This is partly because guidance on the selection of priors is less firm than with other parts of the model. However, there are often strong arguments for particular forms of the prior: little or vague knowledge often justifies a diffuse or even a uniform prior, certain probability models logically lead to particular forms of the prior (conjugacy), and the prior allows researchers to include additional information collected outside the current study. This controversy is also because early leading statisticians were opposed to the use of "subjective" information in specifications, thus this word became a derogatory term for Bayesian practitioners. However, it is a completely misguided charge since *all* decisions made in the process of model development, estimation, and reporting are subjective. It is subjective which data are used; it is subjective which variables are selected; it is subjective which model is selected; it is subjective which likelihood function is specified; it is subjective which significance level is selected (note that the standard levels are not theoretically derived, they are merely "conveniences"); it is subjective which software is used; it is subjective which estimation procedure is specified; it is subjective how the results are presented; and it is subjective where and how findings are distributed. So every statistical model ever developed is "subjective."

There are also technical reasons to allay concerns about the use of prior distributions in statistical models. The maximum likelihood estimate (Section 3) is equal to the Bayesian posterior mode with the *appropriately bounded* uniform prior. This means that a uniform prior can be found to equate the point estimates, although Bayesians generally prefer to give more information than that through distributional descriptions. A likelihood based model with no prior and a corresponding Bayesian model are asymptotically equivalent for any nonpathological choice of prior: as the sample size increases, the likelihood progressively dominates the prior, as demonstrated for one particular case in Section 4.2. Because of the Central Limit Theorem, both estimates are also identically normally distributed for large data. These properties are explored in technical detail in Diaconis and Freedman (1986).

5.1 Conjugate Model Specifications

The phenomenon of conjugacy happens when the distributional form of the prior distribution flows down to the posterior distribution given a specific likelihood function:

$$\pi_f(\theta|\mathbf{y}) = p_f(\theta)L_{\mapsto f}(\theta|\mathbf{y}), \tag{5.1}$$

where "$\mapsto$" means "maps to" and the "f" designation means that $\pi()$ and $p()$ are the same PDF or PMF but with different parameterizations due to the influence of $L()$. We have already seen one example of a conjugate model setup in Section 4 where a gamma distribution prior and a Poisson likelihood produced a gamma distribution posterior distribution. Conjugacy is not a requirement of Bayesian specifications, but it is especially important in the history of Bayesian statistics because before the advent of advanced computational tools in the 1990s it was a mathematically guaranteed way to be able to produce tractable posterior forms. Up until the 1990s it was easy to specify a desired Bayesian (usually regression) model where the joint posterior distribution could not be integrated to produce marginal distributions for a results summary. The most common conjugate relationships are given in Table 3. See Gill (2014), Appendix B for mathematical details on these forms. An excellent book-long exposition that derives these conjugate models and related forms is Zellner (1996).

An easy, obvious, and convenient conjugate setup is the normal-normal for the mean parameter and the normal-inverse gamma for the variance parameter. This is also important because of the popularity of normal specifications in

Table 3 Conjugate prior-likelihood pairings for Bayesian models

Likelihood Function Form	Prior/Posterior Distribution
Bernoulli	Beta
Binomial	Beta
Multinomial	Dirichlet
Negative Binomial	Beta
Poisson	Gamma
Gamma (including χ^2, and Exponential)	Gamma
Normal for μ	Normal
Normal for σ^2	Inverse Gamma
Pareto for α	Gamma
Pareto for β	Pareto
Uniform	Pareto

standard practice and the effect of the Central Limit Theorem. Suppose that we have n-length $\mathbf{y}$, a vector of IID normally distributed data with population mean $-\infty < \mu < \infty$ and population variance $\sigma^2 > 0$. Then the two-variable normal likelihood function for these data is:

$$L(\mu, \sigma^2|\mathbf{y}) = (2\pi\sigma^2)^{-\frac{n}{2}} \prod_{i=1}^{n} \exp\left[-\frac{1}{2\sigma^2}(y_i - \mu)^2\right], \tag{5.2}$$

constructed as shown in Section 3. As per Table 3 we first specify an inverse gamma distribution for the unknown σ^2:

$$\begin{aligned} p(\sigma^2|a, b) &= \frac{b^a}{\Gamma(a)}(\sigma^2)^{-(a+1)} \exp[-b/\sigma^2] \\ &\propto (\sigma^2)^{-(a+1)} \exp\left[-b/\sigma^2\right], \end{aligned} \tag{5.3}$$

where the form in the second line is just the *kernel* of the PDF for σ^2 stripping off the constants with proportionality. If a variable X is distributed gamma, then $1/X$ is distributed inverse gamma, and this distribution is the conjugate form in the normal-normal setup. Here $a > 0$ and $b > 0$ are "hyperparameter values" of the prior that are set by the researcher. An additional utility of this example is in further demonstrating the usefulness of proportionality in simplifying Bayesian calculations. The normal kernel prior distribution for μ is given by:

$$p(\mu|\sigma^2, m) \propto (\sigma^2)^{-\frac{1}{2}} \exp\left[-\frac{1}{2\sigma^2}(\mu - m)^2\right], \tag{5.4}$$

where m is the prior mean parameter for μ set by the researcher. The new wrinkle here is that this prior is actually called a "semi-conjugate" form since it needs to be conditioned on the σ^2 parameter in order for the conjugacy property to hold. The joint posterior distribution is produced in the same way that we did in Section 4, except that there are now two variables instead of one in the model, meaning that the posterior distribution is a "joint posterior" distribution for unknown μ and σ^2 and is thus slightly more complicated than the posterior form in (4.3). This joint posterior distribution is produced from multiplying the two prior distribution kernels times the likelihood function:

$$\pi(\mu, \sigma^2|\mathbf{y}) \propto p(\mu|\sigma^2, m) \quad \times \quad p(\sigma^2|a, b) \quad \times \quad L(\mu, \sigma|\mathbf{y}). \tag{5.5}$$

Now substitute the expressions (5.2), (5.3), and (5.4) into the definition of the posterior distribution earlier and collect like terms into a $(\sigma^2)^{\text{something}}$ component and an exp[something else] component to produce:

$$\pi(\mu, \sigma^2|\mathbf{y}) \propto (\sigma^2)^{-\alpha-\frac{n}{2}-\frac{3}{2}} \exp\left[-\frac{1}{\sigma^2}b - \frac{1}{2\sigma^2}\sum_{i=1}^{n}(y_i - \mu)^2 - \frac{1}{2\sigma^2}(\mu - m)^2\right]. \tag{5.6}$$

This joint form is large and awkward but can easily be marginalized (see details in Gill (2014)) with integration as explained further in Section 6. It is important not to be intimidated by the length of this posterior expression earlier: it is just a series of additive and multiplicative states in two exponentials multiplied together. Performing the marginalization first for σ^2 means integrating out the μ parameter:

$$\begin{aligned}\pi(\sigma^2|\mathbf{y}) &= \int_{-\infty}^{\infty} \pi(\mu, \sigma^2|\mathbf{y})d\mu \\ &\propto (\sigma^2)^{-a-\frac{n}{2}-\frac{3}{2}} \exp\left[-\frac{1}{\sigma^2}\left(b + \frac{1}{2}\sum_{i=1}^{n}y_i^2 - \frac{1}{2}n\bar{\mathbf{y}}^2\right)\right].\end{aligned} \tag{5.7}$$

This feels complicated! The integral averages over uncertainty from the μ parameter leaving just a distributional expression for σ^2, and we will discuss integration extensively in Section 6. Actually this expression earlier is neater than it seems. Looking back at the form of (5.3) we see that this marginal posterior is the kernel of an inverse gamma for σ^2 described by:

$$p(\sigma^2|\mathbf{y}) \propto (\sigma^2)^{-(\text{something}+1)} \times \exp[-(\text{something else})/\sigma^2]$$

where

$$\text{something} = a + \frac{n}{2} + \frac{1}{2}$$
$$\text{something else} = b + \frac{1}{2}\sum_{i=1}^{n} y_i^2 - \frac{1}{2}n\bar{y}^2,$$

which defines the parameters of another inverted gamma distribution! So now we know that the posterior distribution of σ^2 is:

$$\sigma^2|\mathbf{y} \sim \mathcal{IG}\left(a + \frac{n}{2} + \frac{1}{2}, b + \frac{1}{2}\sum_{i=1}^{n} \mathbf{y}_i^2 - \frac{1}{2}n\bar{\mathbf{y}}^2\right). \tag{5.8}$$

Remember that no matter how awkward the parameterization is (although this one is not too bad), once we have the posterior distribution fully described we know everything about it and can describe it to readers any way we want. For instance the mean and variance of the generic inverse gamma PDF for X are given by:

- $E[X] = \frac{b}{a-1}$, $a > 1$.
- $\text{Var}[X] = \frac{b^2}{(a-1)^2(a-2)}$, $a > 2$.

The marginal (really conditional because of the semi-conjugacy) posterior distribution for μ can be obtained by using the definition of conditional probability ($p(A|B) = p(A, B)/p(B)$) and then some basic algebra as follows:

$$\pi(\mu|\sigma, \mathbf{y}) = \frac{\pi(\mu, \sigma^2|\mathbf{y})}{\pi(\sigma^2|\mathbf{y})} = \sigma^{-2}\exp\left[-\frac{n}{2\sigma^2}\left(\mu^2 - 2\frac{n\bar{\mathbf{y}} + m}{n}\mu + \frac{n\bar{\mathbf{y}}^2 + m^2}{n}\right)\right] \tag{5.9}$$

(the $\mathbf{y}$ just "comes along for the ride" here in the conditional probability statement). We can use the same "something" approach as before in looking at this form and see that this is the kernel of another normal distribution, according to:

$$\mu|\sigma^2, \mathbf{y} \sim \mathcal{N}\left[\frac{n\bar{\mathbf{y}} + m}{n}, \frac{\sigma^2}{n}\right]. \tag{5.10}$$

Here that the prior dependence on σ^2 flows through to the posterior for μ, which is why this is called semi-conjugate. Note also that as n gets very large the posterior mean of μ converges to the data mean, which is intuitive but can be shown more technically:

$$\lim_{n\to\infty}\left(\frac{n\bar{\mathbf{y}} + m}{n}\right) = \lim_{n\to\infty}\left(\frac{n}{n}\bar{\mathbf{y}}\right) + \lim_{n\to\infty}\left(\frac{m}{n}\right) = \bar{\mathbf{y}}. \tag{5.11}$$

Table 4 A year 2023 sample of salaries for data scientists in $1,000$s

214.907	125.985	156.206	176.417	210.296	181.129	187.393
172.322	187.450	135.631	162.032	198.629	143.636	188.978
191.123	185.914	153.376	134.740	198.416	138.659	

After splitting up the first fraction into two components we see that the ratio of n values becomes 1 in the first term (even though they are infinity, they are the same "flavor" of infinity), and that the n in the denominator of the second term eliminates that one. This is another clear illustration of the Bayesian principle that the data always "win" in the limit.

5.2 Analysis of Salary Data for Data Scientists

As an example of the normal-normal conjugate Bayesian model, consider the annual US salary numbers in thousands for Data Scientist positions listed for hire at Glassdoor in April 2023 with the sample in Table 4. This is the first 20 listed out of 8,880 since that is the most one can see without a lengthy registration process. Suppose we want to use this modest sample to make inferences about the national picture for Data Science pay. Salary figures in a specific occupation are often approximately normally distributed, even though a small sample of $n = 20$ may not appear to be so. The sample mean is 172.1619 and the sample standard deviation is 26.7349. The hyperparameter values are set at $m = 170, a = 2, b = 550$ to somewhat resemble sample values and provide a diffuse and conservative picture of variability. Calculation of the marginal posterior distributions directly follows from the recipes in (5.8) and (5.10). The results are shown in Figure 6. For the μ parameter we see that the prior is influential in the production of the posterior, which makes sense when the data size is only $n = 20$. This is an important illustration of the sample size principle in Bayesian inference: when the data are modest we will rely heavily on the specification of the prior and should be very introspective about its choice. For the σ^2 parameter we see that the data were more influential in determining the form of the posterior but not dramatically so.

R Code for Normal-Normal Model

```
library(MCMCpack) # FOR dinvgamma(x, shape, scale = 1)
# DATA
salary <- scan("glassdoor.dat")
salary <- salary/1000; n <- length(salary)
```

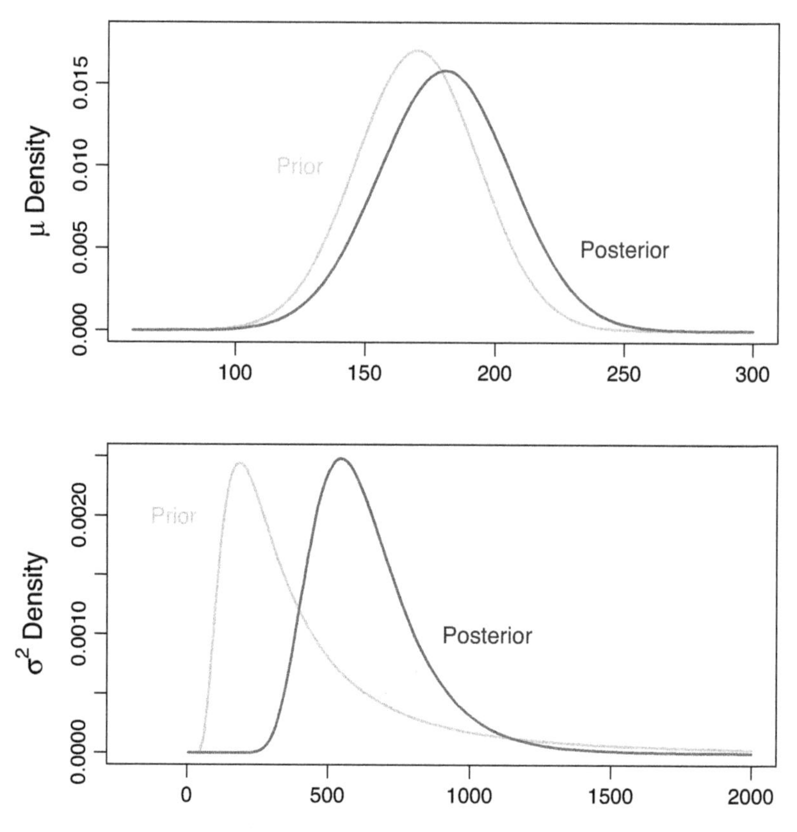

Figure 6 Normal-normal conjugate model for data science salaries

```
# HYPERPRIOR VALUES
m <- 170; a <- 2; b <- 550

# POSTERIOR PARAMETERS
post.a <- a + n/2 + 1/2
post.b <- b + 0.5*sum(salary^2) - 0.5*n*mean(salary)^2
post.mu <- (n*mean(salary) + m)/n
post.var <- post.b/(post.a - 1)

# GRAPH
par(oma=c(1,1,1,1), mar=c(3,5,1,1),cex.lab=1.5,
    mfrow=c(2,1))
ruler <- seq(60,300,length=500)
```

```
prior.dens <- dnorm(ruler,m,sqrt(b/(a-1)))
plot(ruler,prior.dens,type="l",ylim=c(0,0.018),lwd=3,
    col="grey70", ylab=expression(paste(mu," Density")),
    xlab="")
post.dens <- dnorm(x=ruler,mean=post.mu,
    sd=sqrt(post.var))
lines(ruler,post.dens,lwd=3, col="grey30")
text(125,0.010,"Prior",col="grey70",cex=1.10,adj=0.5)
text(250,0.0050,"Posterior",col="grey10",cex=1.10,
    adj=0.5)
ruler <- seq(0,2000,length=500)
prior.dens <- dinvgamma(ruler,a,b)
plot(ruler,prior.dens,type="l",ylim=c(0,0.0025),
    xlim=c(-200,2000), lwd=3, col="grey70", xlab="",
    ylab=expression(paste(sigma^2," Density")))
post.dens <- dinvgamma(ruler,post.a,post.b)
lines(ruler,post.dens,lwd=3,col="grey30")
text(-50,0.0020,"Prior",col="grey70",cex=1.10,adj=0.5)
text(1060,0.0010,"Posterior",col="grey10",cex=1.10,
    adj=0.5)
```

PYTHON Code for Normal-Normal Model

```
import matplotlib.pyplot as plt
import numpy as np
from scipy.stats import norm, invgamma

# DATA
salary = np.loadtxt("glassdoor.dat")
salary /= 1000; n = len(salary)

# HYPERPRIOR VALUES
m = 170; a = 2; b = 550

# POSTERIOR PARAMETERS
post_a = a + n/2 + 1/2
post_b = b + 0.5*sum(salary**2) - \
        0.5*n*np.mean(salary)**2
```

```
post_mu = (n*np.mean(salary) + m)/n
post_var = post_b/(post_a - 1)

# GRAPH
fig, axs = plt.subplots(2, 1, figsize=(8, 12))
ruler = np.linspace(60, 300, num=500)
prior_mu = norm.pdf(ruler, m, scale=np.sqrt(b/(a-1)))
post_mu = norm.pdf(ruler, post_mu, np.sqrt(post_var))
axs[0].plot(ruler, prior_mu, color='grey',
            linewidth=3, label='Prior')
axs[0].plot(ruler, post_mu, color='black',
            linewidth=3, label='Posterior')
axs[0].set_ylabel(r'$\mu$ Density')
axs[0].text(125, 0.010, 'Prior', color='grey',
            fontsize=12, ha='center')
axs[0].text(248, 0.0050, 'Posterior', color='black',
            fontsize=12, ha='center')
ruler = np.linspace(0, 2000, num=500)
prior_s2 = invgamma.pdf(ruler, a, scale=b)
post_s2 = invgamma.pdf(ruler, post_a, scale=post_b)
axs[1].plot(ruler, prior_s2, color='grey',
            linewidth=3, label='Prior')
axs[1].plot(ruler, post_s2, color='black',
            linewidth=3, label='Posterior')
axs[1].set_ylabel(r'$\sigma^2$ Density')
axs[1].text(0, 0.0020, 'Prior', color='grey',
            fontsize=12, ha='center')
axs[1].text(1050, 0.0010, 'Posterior', color='black',
            fontsize=12, ha='center')
plt.subplots_adjust(hspace=0.15)
plt.show()
```

5.3 Typology of Bayesian Priors

Conjugacy is mathematically convenient and leads to easily understood properties, but it is much less important now that we use computational tools to perform the type of operations done in the last section with the normal-normal setup. Modern Bayesians are unencumbered by intractable calculations (marginalization with integration) that were a serious problem throughout most of the twentieth century. So now there exists a wide class of priors to consider, not based on calculability, but on research design criteria.

5.3.1 Proper Bayes

These priors come from previously compiled evidence, such earlier studies or published work, researcher intuition, or substantive experts. It is not unscientific to say that previous research in a specific literature should be considered important prior information in a statistical model. In fact, it is vital to scientific progress to take existing knowledge and improve it systematically as new data are observed. These priors can take whatever distributional form that the researcher considers appropriate, including conjugate forms of course, but typically with regression models normal priors are applied to the regression parameters. These normal forms can be mean centered at: the middle of previous research, skeptically at zero, or with elicited knowledge from experts. Many authors have explicitly argued for the use of such informed priors in the social sciences. Leamer (1972) asserts: "Arguments concerning the use of such prior information should appropriately address the question of how rather than whether prior information should be used." Berk, Western, and Weiss (1995) point out at that "in complex studies where the prior information is based on clearly explained previous studies, the prior may find greater support among skeptical researchers than the model itself." Bartels (1996) uses prior information in a pooling context: "My analysis so far has emphasized that intelligent decisions about how to treat disparate observations must be based, in one way or another, upon prior beliefs about the statistical relevance of the available data."

5.3.2 Empirical Bayes

These prior distributions are produced from other parts of the data, or possibly from the same data used in the likelihood function. One version uses the observed data to establish hyperpriors in a hierarchical model through regular (non-Bayesian) estimation. Another version, called nonparametric empirical Bayes, specifies only a generic form of the prior rather than a specific parametric form. Such approaches leverages both Bayesian and frequentist methods in its implementation and has therefore historically offended both camps. The mid-century Bayesian Lindley famously said "there is no one less Bayesian than an empirical Bayesian" and described the underlying asymptotic justification as "technicalities out of control" (Copas, 1969). However, empirical Bayes estimators enjoy good frequentist properties including a theoretical connection to James-Stein estimation. There are actually many variants as the core ideas have been developing for about 70 years. With the advent of machine learning, interestingly, some of these data re-use ideas have recently become more popular.

5.3.3 Reference Bayes (Objective Bayes)

This approach is really a collection of tools, the most common of which proposes prior distributions that are created to influence the posterior as little as mathematically possible with the idea of being "objective." One way to approach this goal is to specify hierarchical models with objective hyperprior values. To some, any Bayesian procedure is considered objective if it yields good standard frequentist behavior. Often these approaches involve substantial mathematical effort with reasonably complex model specifications, providing a sizable barrier to objective Bayesian work in social science applications.

5.3.4 Decision-Theoretic Bayes

Often setting up decision-theoretic statistical models are difficult to setup in the social sciences because there needs to be a measure of advantage-disadvantage. Economists are blessed with an easy version of this: money. If one can have a measure such as this to evaluate, then Bayesian models conjoin elegantly with the structure of decision theory. Here prior distributions are setup to compare different strategies and results. Results are presented in a full decision-theoretic framework where utility functions determine decision losses, which are minimized according to different probabilistic criteria. See Gill (2014) Chapter 8 for extended details of this approach.

5.3.5 Priors of Convenience

Probably the most common form of prior distributions used by social science Bayesian modelers is a diffuse uniform or normal distribution designed to avoid discussion of prior decisions as much as possible. This is done often more for practical reasons than belief or philosophy: they are easier forms to satisfy skeptical journal reviewers and editors. This approach is both good and bad. In one sense it makes it easier for researchers to pursue Bayesian approaches in various fields and sub-fields that have historically not produced a lot of Bayesian model specifications. On the other hand, it means that researchers are denying themselves the ability to incorporate the deep scientific knowledge that exists all in social science literatures into their model specification.

5.4 Elicited Priors

An important class of proper priors are those that are created from drawing qualitative information out of experts and expressing it in terms of distributional statement. These priors are a way to elicit information from subject-area experts to build a probability structure that captures their specific qualitative knowledge and experience about the phenomenon under study. This elicitation process obtains information from nonstatisticians who have a great deal of contextual information about some substantive issue but are not involved in the model specification process. In published work some of these experts have been physicians, policy-makers, theoretical economists, historians, judges, politicians, previous study participants, outside experts, and community activists.

Elicited priors are often categorized into four general types (Gill and Walker, 2005). *Clinical Priors* are those that are elicited from substantive experts that are taking part in the research project at hand. *Skeptical Priors* are constructed with the assumption that the hypothesized effect does not actually exist and are typically operationalized through a probability function centered at a zero effect. Oppositely, *Enthusiastic Priors* are built around the positions of experts or advocates assuming the existence of the hypothesized effect. *Reference Priors* are produced from expert opinion as a way to express informational uncertainty with diffuse forms (in a somewhat different sense than reference priors discussed earlier).

While there are many specific methods of eliciting priors from experts, these research designs all have essentially three phases. The *Deterministic Phase* starts with specifying explanatory variables in the model and the assumed prior parametric form for their associated coefficients, finding the relevant data collection processes, selecting the number and type of experts to query, and then planning how to evaluate the reliability of their contributions Experts usually need to be trained or briefed before elicitation takes place.

The *Probabilistic Phase* is where experts are actually interviewed is generally the most challenging. There are three common approaches: assessors can be asked fixed value questions with probability responses ("P-methods"), fixed probability questions with value responses ("V-methods"), or questions to be answered on probability and value scales simultaneously ("PV-methods") P-methods determine levels of explanatory variables in advance and ask the assessor to provide the probability of occurrence for different levels of the outcome variable. V-methods ask the more difficult question of determining explanatory variable levels associated with a specific probability value. PV-methods are even more demanding because they require that the assessor simultaneously pick cumulative distribution points and their associated levels as a pair.

Since elicited priors are developed from substantive area experts rather than from the researchers themselves, a key challenge is turning verbal or written opinions into specific probability statements to create prior distributions. A typical strategy is to query experts about outcome variable quantiles for given researcher specified levels of specific explanatory variables. This includes strategies from informal assignments to detailed elicitation plans, as well as regression analysis across multiple experts. Often this process is done online so that assessors fill-in values or select menu options from a formatted window, see the subsequent effects to a visual distribution, and then make adjustments to their original input (see Gill and Freeman (2013)). There are many different specific research designs for elicitation, and this class of prior is a powerful way to bridge qualitative and quantitative knowledge for a given research question.

5.5 Improper Priors

So far all of the prior distributions that we have seen have been "proper," meaning that they conform exactly to the probability rules introduced in Section 2. It is actually possible to specify priors that do not have to conform to all of these rules, in particular the requirement that the PMF or PDF sum or integrate to one. This substantially increases the available selection of distributions that can be specified, particularly when diffuse or skeptical forms are desired. It is important to distinguish between "uninformative" priors such as conjugate forms with large variance, and "improper" priors which are typically uninformative as well but have the mathematical property that they do not sum or integrate to a finite quantity.

Perhaps the most commonly applied improper prior for a parameter that has support $[-\infty:\infty]$ is a uniform (flat) PDF. This is a popular choice for the mean parameter, μ, in a normal model, and is specified by:

$$p(\theta) = c, \quad -\infty < \theta < +\infty. \tag{5.12}$$

for any positive constant c. Think of this as a rectangle that is $H = g(\theta) = c$ high and $W = \infty$ wide. For this form H could be something as simple as $g(\theta) = 1$. It does not seem that this would produce a proper posterior distribution but it turns out that the infinities cancel out in Bayesian inference:

$$\begin{aligned}
\pi(\theta|\mathbf{y}) &= \frac{p(\theta)L(\theta|\mathbf{y})}{\int_\Theta p(\theta)L(\theta|\mathbf{y})d\theta} \\
&\qquad \text{[insert the "box" definition of the prior]} \\
&= \frac{(H \times W)L(\theta|\mathbf{y})}{\int_\Theta (H \times W)L(\theta|\mathbf{y})d\theta} \\
&\qquad \text{[pull } W \text{ out of the integration since it is a constant]}
\end{aligned}$$

$$= \frac{(W)HL(\theta|\mathbf{y})}{(W)\int_\Theta HL(\theta|\mathbf{y})d\theta}$$

[cancel the Ws since they are the same flavors of ∞]

$$= \frac{HL(\theta|\mathbf{y})}{\int_\Theta HL(\theta|\mathbf{y})d\theta}$$

[substitute back $g(\theta)$ for generality]

$$= \frac{g(\theta)L(\theta|\mathbf{y})}{\int_\Theta g(\theta)L(\theta|\mathbf{y})d\theta},$$

resulting in a proper posterior statement like the definitional statement in (4.3). The cancellation is possible because mathematicians tell us that this can be done with the same "flavor" of infinity, as opposed to infinities that are produced from distinct processes: $f(\theta, \infty) \neq g(\theta, \infty)$. Here these two infinities are produced from the exact same specification so the cancellation works.

There are also some popular improper choices for the variance parameter in the normal model: $p(\sigma) = 1/\sigma$, and $p(\sigma) = k$, for some constant k (these forms are usually given for σ not σ^2, but this is not important conceptually). The logic of how these work is the same as earlier. Compared to conjugate forms, improper forms always lead to posterior distributions with larger variance. This is intuitive since there is less information in an improper prior. For instance, recall that a normal-normal conjugate setup leads to a normal posterior for μ, but an improper prior with a normal likelihood produces a Students-t posterior distribution, which has heavier tails than a normal and therefore greater variance. More specifically, stipulate a model to estimate μ and σ with a normal likelihood and the two improper priors:

$$p(\mu) \propto c, \quad -\infty < \mu < \infty$$
$$p(\sigma) \propto \sigma^{-1}, \quad 0 < \sigma < \infty,$$

(for some positive constant c) with the goal of adding very little prior information. Then the marginal posterior distribution for μ is Students-t, which is most intuitively described by a transformation:

$$\frac{\mu - \bar{\mathbf{y}}}{s/\sqrt{n}} \sim t(df = n-1),$$

where $s = \sqrt{(n-1)^{-1}\sum(y_i - \bar{\mathbf{y}})^2}$. So the marginal posterior of μ is also Student's-t with noncentrality parameter $\bar{x}$, thus providing a more diffuse than the normal conjugate form since Students-t distributions have heavier tails. Also, the resulting marginal posterior for σ^2 is $\mathcal{IG}((n-2)/2, (n-1)s^2)$ is generally more diffuse version than the form we saw before for the conjugate case in (5.8), depending on the choices for the a and b parameters therein.

5.6 General Thoughts on Priors

At this point it is important to note that there is no such thing as "noninformative" prior distributions. For reasonable sized finite samples any specified prior has an influence on the posterior distribution, even if small. A better term is "uninformative" when the prior has little influence over the posterior. There was an acrimonious debate about whether uniform priors were actually noninformative that took a surprisingly long time to resolve. Even in the world of Reference Bayesians there is recognition that all priors matter with finite samples. The core of this issue is that stating that discrete alternatives or regions of continuously measured space are equally likely turns out the be an informed statement. Consider three candidates for election to US president: a Democrat, a Republican, and a Communist, where all are on the ballot. In any electoral context it is a ridiculous, and nonconservative in the modeling sense, statement to say that they are all a priori equally likely to win.

An enduring criticism that has little validity is the idea that one can fix the prior distribution in a strategic way to shape the posterior distribution as one wants. Stipulating priors is an overt public statement made to a naturally skeptical scientific audience, so priors cannot be covertly adjusted to "cook" the conclusions. This is where the "subjective" term has been wrongly used in the past. Specification of prior distributions should be clearly stated and defended along with any other model decisions made by researchers.

There is another useful aspect of prior distributions. When there are multiple theories or empirical observations, then the associated prior distributions can be used to test the efficacy of the different subsequent posterior distributions. Producing even multiple priors is a straightforward and easy process and can be used to test alternative theories about the state of the world. See Wagner and Gill (2005) for an example.

Finally, in this section we note that the idea of prior information, and even prior distributions, is not restricted to specifically Bayesian modeling. Scholars frequently talk about prior theories, prior findings, and prior conclusions. It is also common to find authors discussing results in overtly prior-posterior terms in published work. For instance, Canes-Wrone, Brady, and Cogan (2002) (clearly not Bayesian practitioners) in discussing model results state:

> The effects of campaign spending and district ideology are consistently in the expected direction and statistically significant. Those on challenger quality also have the correct sign in each regime and sample and are significant with the exception of the marginal regime of the 1980-1996 test. In addition, the coefficients for the remaining variables that are not included as a main

effect typically have the predicted sign, and they are significant only with the expected sign.

There are at least four prior-posterior assertions in those sentences!

6 Integrals and Expected Value: Not as Scary as they Look

The development of calculus in the seventeenth century by Newton and Leibniz is considered a turning point not just in mathematics but also in human development. The credit here is a bit of a simplification as other mathematicians of the time made important contributions as well. Basic calculus is now offered as a course at many high schools and is considered essential knowledge in many fields. The central principle in calculus is determining what happens to functions in limits ($\rightarrow \infty$) and infinitesimals ($\rightarrow 0$). In this section we introduce the use of integrals which are central to Bayesian inference, and statistics in general. To restate a point in the introduction in starker terms: *the required use of integrals in Bayesian statistics is the main reason that students and other researchers cannot easily slide from an introductory statistics course to learning Bayesian statistics.* This section is the linchpin of our strategy to rectify this problem. Since computers are used to perform the actual analytical work, the emphasis here is on general principles to interpret the output of these operations rather than the mathematical details.

6.1 Expected Value for Discrete Random Variables

Expectation can be thought of as a mean average taken over a random (stochastic) target rather than just a vector of numbers. Returning to flipping a coin, consider it now as a gambling exercise. The probability of a heads for a given, not necessarily fair, coin is $p(heads) = a$, where a is a number between zero and one as described in section 2. Therefore the probability of tails is given by $p(tails) = 1 - a$. Suppose you are offered the opportunity to flip this coin where you would get \$1 if it comes up heads and \$0 if it comes up tails. What is this game worth? Meaning, what would you be willing to pay to play this game. If $a = 0.5$ for a fair coin you might feel that it is worth playing this game if it costs you less than 50¢, especially if you could play multiple times. Why is that? It is because over time your expected earnings are positive. Consider the payout and probability for each of the two outcomes in a single toss where the game costs 49¢ to play:

- Heads: cost = −0.49¢, payout = 0.5× \$1
- Tails: cost = −0.49¢, payout = 0.5× \$0,

since there is a 50% chance of either outcome. The *expected value* of playing this particular game is therefore:

$$EV(flip) = -0.49 + (0.5 \times 1) + (0.5 \times 0) = 0.01, \tag{6.1}$$

meaning that the expected value is one penny. It doesn't mean that you are going to win 1¢ on any play of the game: you will win either \$0 or \$1. It means that over a long sequence of plays your average (expected) return per game is 1¢. The expected value here would have been 50¢ if there was not a cost to play this game (the more general case), meaning that the expected value is 0.5 times \$1 plus 0.5 times \$0. *Thus expected value is a balance between chance and reward.* Clearly you would not want to play this game if it costs 99¢ for each round. This exact idea is how casinos make money. Every game in a casino has a positive expected value to the house and a negative expected value to the player, although this number varies considerably across the various games offered.

To further make this balance between chance and reward more intuitive suppose you were offered \$100 to walk the length of a 2×4 lumber board sitting on the ground. It seems correct to say there is a probability of about 0.99 that you would successfully walk on the board from one side to the other and therefore a probability of about 0.01 that you would stumble and step off the board on to the ground 2 inches below with no obvious consequences. You would certainly want to play this game since:

$$EV(walk) = 0.99 \times 100 + 0.01 \times 0 = 99,$$

meaning that the expected payout is \$99. Now suppose we take this exact same board and we put it 40 feet in the air across two adjacent buildings. Barring any psychological issues the probability of walking across the board is still the same at 0.99, and the payout is still \$100. What is different? Now the cost of stumbling off the board is no longer near zero since 40 feet is the height that kills half of people falling ("LD50" in epidemiologist terms) . So most people would probably see the cost of stumbling now as $-\infty$. This makes the expected value calculation:

$$EV(walk) = 0.99 \times 100 + 0.01 \times -\infty = -\infty,$$

where the minus sign is because this is now a penalty not a different reward. So you would definitely not want to play this version of the game. What changed was not the probability of either event but what you get for one of the outcomes. Oddly, there is almost certainly dollar amount reward that would entice *some* people to play. A million dollars? Ten million dollars? We can slightly formalize this logic in order to conceive new board-walking games for the arbitrary outcome a:

$$EV(walk) = p(a) \times reward(a) + p(not\ a) \times reward(not\ a),$$

where a is the event of remaining on the board, or some other event that we could stipulate. Casinos make (a lot of) money by manipulating these four components earlier in a way that exploits human psychology. Apparently humans are poor assessors of events with very high or very low probabilities, which is why lotteries exist.

Now let us take this idea and make it more statistical in notation. Start with a discrete random variable Y with K possible outcomes. These outcomes are analogous to the rewards/punishments idea earlier, but they are numerical, like $heads = 1$, $tails = 0$. This is described by the Bernoulli PMF $p(Y = k) = p_k$, $k = 1, 2, \ldots, K$, with the requirement that $p_1 + p_2 + \ldots + p_K = 1$. If $k = 2$ then we would have the coin flip or the board walking setup. We can state this more formally as the expected value of the discrete random variable Y:

$$E[Y] = \sum_{k=1}^{K} p(y_k) y_k \tag{6.2}$$

meaning that the expected value is the sum of: the values of each event times its corresponding probability. This expected value calculation can be applied to any PMF. For example, consider a binomial experiment of n trials with y successes:

$$p(Y = y|n, p) = \binom{n}{y} p^y (1-p)^{n-y}, \quad n \geq y,\ n, y \in \mathcal{I}^+,\ p \in [0{:}1]. \tag{6.3}$$

where $\binom{n}{y} = \frac{n!}{y!(n-y)!}$. The expected value of this random variable is then:

$$E[Y] = \sum_{i=1}^{n} p(y_i) y_i = E[y_1] + E[y_2] + \ldots + E[y_n] = np \tag{6.4}$$

since there are n trials each with identical probability p. Here we used the expected value relation: $E[X + Y] = E[X] + E[Y]$, which a property that actually applies to both sums and integrals.

As a quick example of such a calculation LeBron James' lifetime NBA free throw percentage is 73.5% as of April 2023, which is an expected value of taking a single attempt. So if he takes 15 free throws in a given game we would *expect* approximately 11 baskets ($0.735 \times 15 = 11.025$ rounded down).

6.2 Summing over Continuous Space

Suppose we wanted to measure the area under a curve or function $f(x)$ on the real line, $\mathfrak{R}$, such as the one depicted in Figure 7. As seen in the figure one way to do this is to use a histogram, where in this case the upper left corner of the bars touches the curve. It could be the right; it doesn't matter. This histogram estimate is the sum of the area of the B bars with the same width but different heights:

$$A = \sum_{B=bars} (\text{height}_{bar} \times \text{width}_{bar}). \tag{6.5}$$

The first panel the histogram has $B = 25$ bars and as a result when the curve is increasing it underestimates the area and when the curve is decreasing it over estimates the area. This would have been the other way around if we used the upper right corner to touch the curve instead. The total error is the sum of the

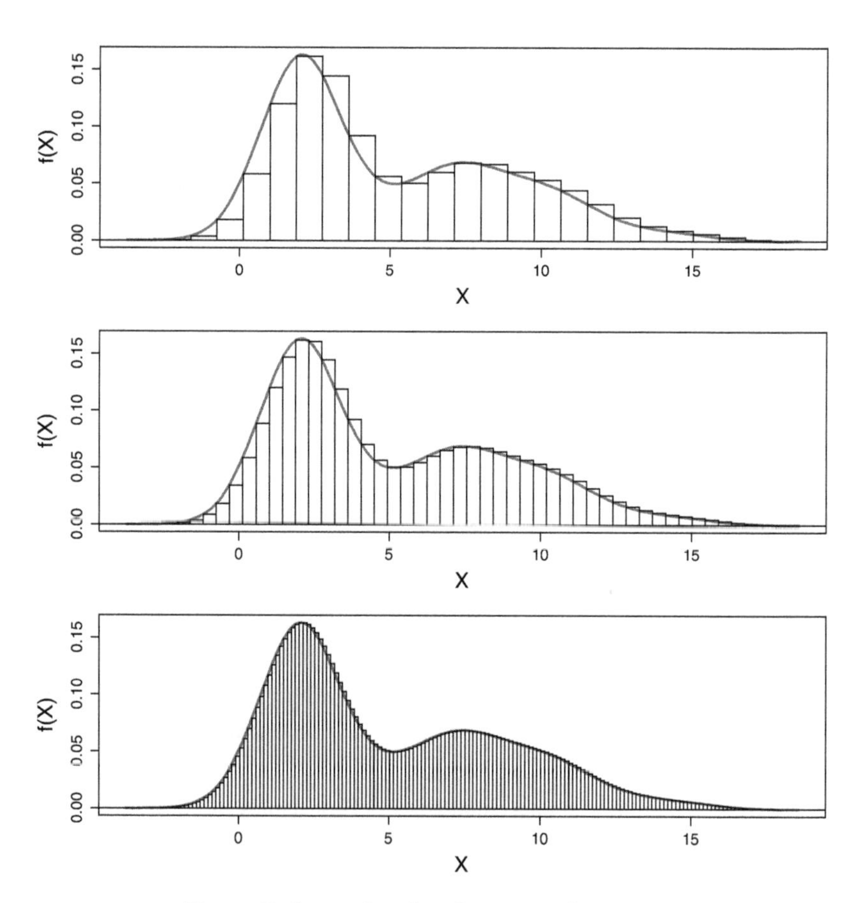

Figure 7 Approximating the area under a curve

area in these under- and over-counts: the areas between the histogram and the curve summed. In the second panel there are $B = 50$ bars in the histogram, and one can see that the errors in estimating the area are smaller. In the third panel there are $B = 164$ bars, and it is clear that now the errors are smaller still. What would happen if we took $B \longrightarrow \infty$? It is intuitive that there would then be no errors at all and the sum of the area of the bars would exactly equal the area under the curve. This is precisely how an integral works in calculus: set the area criteria (as we have done with the histogram in this case) and then take a discrete approximation method to the limit. In notation this looks like:

$$A = \int_a^b f(x)dx \tag{6.6}$$

where the limits a and b are the starting and ending points for the area of interest on the x-axis (real line, $\Re$). We first saw an expression like this in (4.2). As noted before, "$\int$" is a smooth version of Σ since we are on the real line instead of across discrete events, and "dx" is a reminder that the integration function is with regard to the variable x.

Notice that the integration process starts with a function, a distribution function in our case, and ends with a number. Thus the uncertainty inherent in a PDF is quantified as a single number. For example if we integrated a standard normal PDF, $f(x) = (2\pi)^{-1} \exp[-x^2]$, over its complete support, $[-\infty : \infty]$, then this operation would return the number 1 by definition of a PDF. If we integrated this same standard normal over $[0 : \infty]$ then we would get the number 0.5. This second integration means that we have integrated half the area under the curve of the PDF and described half of the uncertainty.

Before in Section 4 we needed to use integrals to introduce key elements of Bayesian inference and integrals in this process and were discussed more intuitively rather than mathematically. Now that we have a theoretical definition, we can return to these operations in slightly more detail. One important way that integrals are used is to take a joint distribution between two interval measured variables and integrate out one of them to produce a marginal distribution for the other. In abstract terms it looks like this:

$$f(x) = \int_y f(x,y)dy,$$

where the integration is over y so we are "integrating out uncertainty in y" from the joint distribution of x and y to produce a marginal distribution for x. By using just y under the integral sign and "dy" we are stating that the integration takes place over the complete support (range) of this variable. So the first integral that we saw in Section 4 took place over a joint function of θ and $\mathbf{y}$ given by $p(\theta)L(\theta|\mathbf{y})$. Integrating θ over its complete support Θ gave the marginal distribution of $\mathbf{y}$:

$$\int_{\Theta} p(\theta)L(\theta|\mathbf{y})d\theta = f(\mathbf{y}).$$

This turned out not to be a distribution by the standard Bayesian assumption that once the data are observed they are fixed so $f(\mathbf{y})$ is just a constant. Then in (4.3) we used this integral-produced value to make sure that the posterior probability for θ was a proper probability statement with the criteria given in Section 2 by normalizing the prior times the likelihood in the numerator:

$$\pi(\theta|\mathbf{y}) = \frac{p(\theta)L(\theta|\mathbf{y})}{\int_{\Theta} p(\theta)L(\theta|\mathbf{y})d\theta}.$$

Later in the exposition of the normal-normal conjugate model the integral in (5.7) was used to integrate out the μ parameter from the joint posterior distribution to give the marginal posterior distribution of σ^2:

$$\int_{-\infty}^{\infty} \pi(\mu, \sigma^2|\mathbf{y})d\mu = \pi(\sigma^2|\mathbf{y}).$$

So integration is used to describe the area under a curve, but all of our "curves" are PDFs describing probability density over defined intervals on the real line (including those going to infinity like the normal). Thus when we integrate an interval-measured random variable over its support we get the number 1, and when we integrate out an interval measured random variable from a joint probability statement we are integrating out the uncertainty contained in this probability statement. Furthermore, this process is not constrained to the treatment of just two random quantities. For instance:

$$\int_{\Theta} p(\alpha, \beta, \theta|\mathbf{y})d\theta = p(\alpha, \beta|\mathbf{y}).$$

Obviously there are more complex versions of this principle, but they all have the same characteristic of integrating out the uncertainty contained in a specific target variable or variables.

6.3 Expected Value for Interval Measured Random Variables

After discussing expected value for discrete random variables and understanding the basic principles of integration we can now move on to expected value for interval measured random variables, which is the most common case in Bayesian inference. Since we have done all of this preparatory work this section will be short.

This interval measured expected value is nearly identical to (6.2), except that the summation is replaced with integration:

$$E[Y] = \int_{-\infty}^{\infty} Yp(Y)dy,$$

here the limits are given as infinities just as a matter of generality: we use finite bounds if appropriate or stipulate the bounds by identifying the support of the integrand as in (4.2). Since most random variables (not data) that we work with in Bayesian models are interval measured, this is the most important form of the expected value.

As stated before, this expected value is mean of the distribution of the variable of interest as used in (4.10) from the known mean $E[X] = \frac{\alpha}{\beta}$ (rate version) of a gamma PDF. But we now have the tools to prove that this is true as an example of working with interval measured expected value. Starting with (6.3) and the definition of the gamma PDF for a random variable X:

$$\begin{aligned}
E[X] &= \int_0^{\infty} (X)\beta^{\alpha}\Gamma(\alpha)^{-1}x^{\alpha-1}\exp[-\beta X]dx \\
&\qquad \text{[collect exponents of } x\text{]} \\
&= \int_0^{\infty} \beta^{\alpha}\Gamma(\alpha)^{-1}x^{\alpha+1-1}\exp[-\beta X]dx \\
&\qquad \text{[pull } 1/\beta \text{ out of the first term]} \\
&= \int_0^{\infty} \frac{1}{\beta}\beta^{\alpha+1}\Gamma(\alpha)^{-1}x^{\alpha+1-1}\exp[-\beta X]dx \\
&\qquad \text{[use the relation } \Gamma(x+1) = x\Gamma(x)\text{]} \\
&= \int_0^{\infty} \frac{\alpha}{\beta}\beta^{\alpha+1}\Gamma(\alpha+1)^{-1}x^{\alpha+1-1}\exp[-\beta X]dx \\
&\qquad \text{[move } \alpha/\beta \text{ out of the integral since it is a constant]} \\
&= \frac{\alpha}{\beta}\int_0^{\infty} \underbrace{\beta^{\alpha+1}\Gamma(\alpha+1)^{-1}x^{\alpha+1-1}\exp[-\beta X]}_{\mathcal{G}(x|\alpha+1,\beta)}dx \\
&\qquad \text{[now integrating a different Gamma PDF over its full range]} \\
&= \frac{\alpha}{\beta} \times 1
\end{aligned}$$

Note the manipulation of the Gamma function (the noninteger version of the factorial function given in Section 4.2). We were also able to pull out α/β out of the integral calculation since it is a constant ratio of two numbers and therefore is not affected by the integration process, for an arbitrary constant k: $\int kf(x)dx = k\int f(x)dx$. We can also derive the expected value of a random variable distributed inverse gamma given in Section 5.1, $E[X] = \frac{b}{a-1}$, $a > 1$, by similar processes.

Consider how we would use simulation to get the expected value of a distribution where we could not simply look up this answer. Figure 7 was created from an even (50-50) mixture of two normal distributions according to:

$$p(y|\boldsymbol{\pi}, \mu_1, \sigma_1^2, \mu_2, \sigma_2^2) = \pi_1\phi_1(y|\mu_1, \sigma_1^2) + \pi_2\phi_2(y|\mu_2, \sigma_2^2), \tag{6.7}$$

where ϕ denotes the normal PDF as in the likelihood form of (5.2) and the proportional prior in (5.4):

$$\phi(y|\mu, \sigma^2) = (2\pi\sigma^2)^{-\frac{1}{2}} \exp\left[-\frac{1}{2\sigma^2}(y-\mu)^2\right], \tag{6.8}$$

and $\boldsymbol{\pi}$ is vector of weights containing only two values: (0.5, 0.5). For the figure we used $\phi_1(y|2, 1)$ and $\phi_2(8, 3)$. Obviously the expected value of this mixture distribution is an evenly weighted mean of the μ parameters based on the property of weights and the structure of the normal PDF. Now suppose we are interested in calculating the expected value for a different mixture of distributions, $\mathcal{N}(10, 10)$ and $\mathcal{G}(4, 6)$, with the pair of weights: $\pi = (0.15, 0.85)$. There are several ways to get the expected value but simulation is perhaps the fastest and easiest. This is merely two lines of R or Python code (or one line to be more clever) as shown in the adjacent code boxes.

R Code for Mixture Distribution Expected Value

```
y <- c(rnorm(15000,10,10), rgamma(85000,shape=4,
    scale=6))
mean(y)
```

Python Code for Mixture Distribution Expected Value

```
import numpy as np

y = np.concatenate((
    np.random.normal(loc=10,scale=10,size=15000),
    np.random.gamma(shape=4,scale=6,size=85000)))
np.mean(y)
```

7 Software Calculation of Bayesian Models

In previous sections, we have explored the fundamental principles of Bayesian statistics and discussed how to construct Bayesian models. Nevertheless, in practice, it is rare to perform these calculations manually. With the dramatic increase in computing power over the past few decades, along with the development of robust software packages designed to handle intricate computations, Bayesian analysis has become increasingly accessible. This section aims to introduce some of the commonly used software tools for constructing Bayesian models. The goal is to provide an overview of the range of software and programming tools. This information will help extend the study of Bayesian statistics beyond the material covered in this Element. And by having a sense of the tools available, readers will be better prepared to tackle more complex problems.

7.1 Basic Functions: Probability and Simulation Functions

Probability distributions and simulations play a crucial role in Bayesian statistics. As discussed extensively in the previous sections, both prior and posterior distributions are expressed in terms of probability. The likelihood function that bridges the two distributions, is also technically a product of the PMF or PDF of the observed data. Furthermore, going deep into the Bayesian realm, the calculation of posterior distributions often doesn't have a simple, closed-form solution. This is where simulation methods, primarily Markov chain Monte Carlo (MCMC), are used to draw samples from the posterior distribution and estimate quantities of interest.

Here we start with some basic functions that are frequently used in the analysis. These functions include calculating probabilities, manipulating distributions,and simulating data. Knowing these functions is important to understand the mechanics of Bayesian models. Moreover, these functions would also facilitate the functionalities of more advanced tools later on.

In the earlier sections, we have used `rnorm` to generate random values from the normal distribution (Section 2) and `dpois()`, `dnorm()`, and `dinvgamma` to generate prior and posterior densities of poisson, normal, and inverse gamma distributions (Section 5.2). The adjacent text boxes provide some common PMF and PDF functions in `R` and `Python`. For `R`, base `R` provides some most fundamental distributions, and additional distributional functions are provided by other non-base packages.

`R` Functions for Computing the PDFs of Function Forms

```
dnorm(x, mean = 0, sd = 1)
dbinom(x, size, prob)
dbeta(x, shape1, shape2)
```

```
dnbinom(x, size, prob, mu)
dpois(x, lambda)
dunif(x, min = 0, max = 1)
dcauchy(x, location = 0, scale = 1)
dchisq(x, df, ncp = 0)
dexp(x, rate = 1)
```

R Functions Provided by Non-Base Packages

```
dexp(x, rate = 1) dinvgamma(x, shape, rate=1)
    #MCMCpack PACKAGE
ddirichlet(x, alpha) #gtools PACKAGE
dmvnorm(x, mean, Cov) #mvtnorm PACKAGE
dinvchisq(x, nu, tau) #extraDistr PACKAGE
dlaplace(x, mean = 0, sd = 1) #jmuOutlier PACKAGE
dpareto(x, location, shape = 1) #EnvStats PACKAGE
diwish(W,v,S) #MCMCpack PACKAGE
```

These distributional functions also usually come with different forms of functions to calculate CDF and quantiles, and generate random numbers following the distributions. Using the normal distribution as an example:

- `dnorm(x, mean, sd)`: Evaluates the PDF of the normal distribution at the specified values of x with given mean and standard deviation.
- `pnorm(x, mean, sd)`: Computes the CDF of the normal distribution up to the values of x with given mean and standard deviation.
- `qnorm(p, mean, sd)`: Calculates the quantiles of the normal distribution corresponding to the given probabilities p with specified mean and standard deviation.
- `rnorm(n, mean, sd)`: Generates n random numbers following the normal distribution with given mean and standard deviation.

For `Python`, distribution functions are usually provided through libraries, such as `SciPy` and `NumPy`. `NumPy` is usually used for random number generation based on distributions while `SciPy` provides a wide range of distribution functions, including probability density functions (PDFs) and cumulative distribution functions (CDFs).

PYTHON Functions for Computing the PDFs of Function Forms

```
scipy.stats.norm.pdf(x,loc=0,scale=1)
scipy.stats.binom.pmf(x,size,prob)
```

```
scipy.stats.beta.pdf(x,shape1,shape2)
scipy.stats.nbinom.pmf(x,size,1 - prob,mu)
scipy.stats.poisson.pmf(x,lambda_)
scipy.stats.uniform.pdf(x,loc=0,scale=1)
scipy.stats.cauchy.pdf(x,loc=location,scale=scale)
scipy.stats.chi2.pdf(x,df,loc=0,scale=1)
scipy.stats.expon.pdf(x,loc=0,scale=1)
scipy.stats.gamma.pdf(x,shape,scale=1)
scipy.stats.dirichlet.pdf(x,alpha)
scipy.stats.multivariate_normal.pdf(x,mean,Cov)
scipy.stats.invgamma.pdf(x,a,loc=0,scale=1)
scipy.stats.laplace.pdf(x,loc=0,scale=1)
scipy.stats.pareto.pdfpdf(x,b,loc=0,scale=1)
```

PYTHON Code for Other Functions of the Normal Distribution

```
scipy.stats.norm.pdf(x,loc=0,scale=1) #PDF
scipy.stats.norm.logpdf(x,loc=0,scale=1) #Log of PDF
scipy.stats.norm.cdf(x,loc=0,scale=1) #CDF
scipy.stats.norm.logcdf(x,loc=0,scale=1) #Log of CDF
scipy.stats.norm.ppf(q,loc=0,scale=1) #Quantile
scipy.stats.norm.rvs(loc=0,scale=1,size=1) #Random number
numpy.random.normal(loc=0.0,scale=1.0,size) #with NumPy
```

It is also worth mentioning that, in `Python`, `tensorflow_probability` is one of the most powerful and comprehensive library that contains a wide range of probability distributions in flexible forms. The adjacent code box replicates the example in Section 2:

PYTHON Code for Probability with TensorFlow

```
import tensorflow as tf
import tensorflow_probability as tfp
tfd = tfp.distributions
n_sims = 1000000
norm_dist = tfd.Normal(loc=3., scale=2.)
y = norm_dist.sample(n_sims)
tf.reduce_mean(tf.cast(y > 0, tf.float32))
# CDF FOR PROBABILITY > 0
(1 - norm_dist.cdf(0)).numpy()
```

These functions that work with various distributions allow simulation analysis like in Section 2 as well as direct querying of distributional properties. Since Bayesian inference is fundamentally about probability distributions the use of these tools is important for doing even basic work.

7.2 Bayesian Packages and Probability Programming

For most of the time, we do not need to construct Bayesian models from scratch like we are doing in most parts of this Element (for pedagogical purposes). Instead, we can take advantage of software packages and libraries that offer ready-to-use functions and tools either specifically designed for Bayesian analysis or for general probability programming that can be used to implement Bayesian models. These tools provide efficient algorithms and implementations for posterior inference, parameter estimation, model fitting, and so on. The following section provides a brief overview of the different types of tools. It is important to note that the full range of these tools may exceed the scope of this Element. Additionally, various tools often can achieve the same functionality. As a result, users typically only need to choose one that best suits their needs. Nonetheless, by exploring these tools, we can have a practical understanding of their various functionalities and capabilities and make informed choices when progressing toward the stage of advanced Bayesian modeling.

These packages and libraries can be roughly grouped into four main categories: (1) Bayesian specific `R` packages or `Python` libraries, (2) Bayesian probabilistic programming platforms, (3) general probabilistic programming framework, (4) general-purpose programming language.

Table 5 lists some example Bayesian packages/libraries in `R` and `Python`. Packages/libraries, such as `MCMCpack`, `pymc`, `nimble`, and `brms`, enable users to easily perform Bayesian computations without having to build Bayesian models from scratch. `JAGS` and `Stan` are standalone Bayesian programs/languages that perform Bayesian analysis but are also available in R and Python through interfaces, for example, `rjags`, `pyjags` (for `JAGS`) and `rstan`, `pystan` (for `Stan`). Note that although `brms` and `rstanarm` also uses `Stan` in the back-end, they allow users to use simple `R` modeling syntax to construct models without knowing the `Stan` syntax. Since `Python` is a general-purpose programming language (compared to `R` as a statistical computing focused language), therc are more general probabilistic programming libraries providing modules for probabilistic modeling and inference, including `pyro` and `tensorflow_probability`, which can also be used to build Bayesian models and perform Bayesian computations leveraging PyTorch's and TensorFlow's computational capabilities, respectively. In the context of `R`, for more

Table 5 Bayesian packages and libraries in R and Python

R/Python **Interfaces**	**Back-end**	**Written in**
MCMCpack, bayesm	-	C++
pymc	aesara, JAX	Python
brms, rstanarm	Stan	C++
nimble	JAGS	C++, R
emcee	-	Python
rjags, pyjags	JAGS	C, Fortran
rstan, pystan	Stan	C++
pyro*	pyTorch	C++, Python
tensorflow_probability*	TensorFlow	C++, Python
Rcpp*	C++	-

Note: * pyro, tensorflow_probability, and Rcpp are not specifically Bayesian packages/libraries.

customized and efficient options, advanced users may need to use Rcpp as an interface and write tailored Bayesian algorithms in C++ to construct more complex Bayesian models. Technically, this is not very different from building a model from scratch in other general-purpose languages including R or Python. However, C++ as a lower-level programming language can provide significant improvements in performance and efficiency while Rcpp interface still allows users to access the data handling capabilities of R, therefore becoming a popular choice for specifying advanced, highly customized Bayesian models.

Bayesian-specific packages/libraries tend to provide ready-to-use Bayesian modeling functionalities with default setups and presupplied Bayesian algorithms. They tend to limit the modeling options to what the packages/libraries currently provide. As an example, we can replicate the code in Section 5.2 using MCMCpack and PyMC3 as shown in the adjacent code boxes.

R Code for Normal-Normal Model with MCMCpack

```
library(MCMCpack)
# DATA
salary <- scan("glassdoor.dat")
salary <- salary/1000
# HYPERPRIOR VALUES
```

```
m <- 170; a <- 2; b <- 550
# POSTERIOR SAMPLES
# MCMCregress USES c0/2, d0/2 FOR INV. GAMMA PARAMETERS
posterior_samples <- MCMCregress(salary ~ 1,
                                 n.samples = 10000,
                                 b0 = m, B0 = 1,
                                 c0 = a*2, d0 = b*2)
post.mu <- mean(posterior_samples[,1])
post.var <- mean(posterior_samples[,2])
```

PYTHON Code for Normal-Normal Model with PyMC

```
import numpy as np
import pymc as pm
m = 170; a = 2; b = 550
# DEFINING MODEL & SAMPLING
with pm.Model() as model:
    s_sq = pm.InverseGamma('sigma_sq', alpha=a, beta=b)
    mu = pm.Normal('mu', mu=m)
    obs = pm.Normal('obs', mu=mu,
                    tau=1/s_sq, observed=salary)
    trace = pm.sample(10000, tune=1000, chains=3)
post_mu = np.mean(trace.posterior["mu"])
post_var = np.mean(trace.posterior["sigma_sq"])
```

For standalone Bayesian programs such as JAGS and Stan, the primary advantages are flexibility and their robust support for Bayesian calculations. They provide a wide range of built-in distributions and functions and allow users to have enough control over the model–fitting any model JAGS and Stan can handle. However, although we can still interact them through R or Python, they require different syntax.

These Bayesian packages/libraries and probability programming platforms provide capabilities for constructing advanced and complex Bayesian models. However, it is important to recognize that there are oftentimes tradeoffs between ease of use and flexibility. Packages like MCMCpack, pymc, and brms provide user-friendly interfaces with simple and streamlined workflows, but they have limitations in terms of model complexity and customization. In contrast, frameworks such as JAGS, Stan, and TensorFlow provide more general and flexible solutions while requiring a steeper learning curve of the specific syntax and configuration, and also a deeper understanding of the underlying concepts.

8 Evaluating and Comparing Model Results

In this section we describe different ways to present results from Bayesian inference. As noted, Bayesians not only have the *ability* to describe inferential outcomes with more information than a point estimate and curvature around it, but they *prefer* to do this since this it gives readers a more nuanced and informed picture of how information has been updated. Results can be described more completely because of the central tenet of Bayesian inference: all unknowns are given probability statements (distributions or values). Therefore the posterior distribution at the end of the modeling process can be described with any feature of a distribution that highlights important results and findings. These are statements that non-Bayesians are incapable of making. In fact an attempt to do so by a non-Bayesian would actually make them a Bayesian at that point!

Consider the normally distributed posterior distribution for the mean parameter μ depicted in Figure 8 given by $\mathcal{N}(3.1695, 2)$. In this depiction the proportion of the density to the left of zero is 0.06 and the proportion of density to the right of zero is 0.94. If this were part of a non-Bayesian one-sided hypothesis test with a normal assumption we would fail to reject the hypothesis that $\mu > 0$ in the population at the ubiquitous $\alpha = 0.05$. (Proclaimers of the importance of $\alpha = 0.10$ are usually those that didn't meet the first threshold in that world.) Conversely a Bayesian looks at this posterior distribution and can say "there is a 94% probability that the effect is positive." Would you bet your own money that this is truly a positive effect? Almost everyone would. Fortunately nobody actually has to bet money. It is a strong evidential statement on its own to say that our inference implies that $p(\mu > 0) = 0.94$ in the population given the sample and the model. Readers can readily assess such statements with regard to the quality of evidence.

It is critical to note again that non-Bayesians cannot make such a probabilistic assertion as done earlier, and those researchers in the social sciences and elsewhere often revert to the contorted (and wrong) logic of the Null Hypothesis Significance Test (NHST) (Gill, 1999). This approach to hypothesis testing has long dominated basic statistical decision-making despite literally thousands of published articles across many scientific fields noting its deep and critical flaws. The worst feature is that it is built on a logical paradigm that fails to hold: if the null hypothesis of no effect is true then the data are highly likely to follow an expected pattern; the data do not follow this expected pattern because they produce a test statistic in the tail of the null distribution implied by this

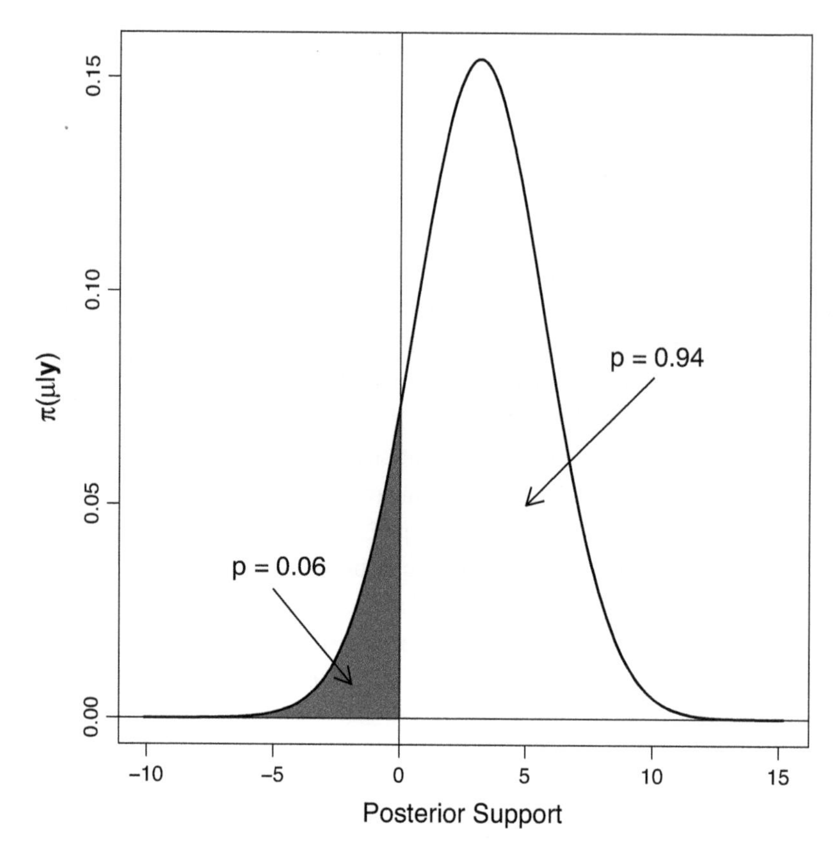

Figure 8 Areas under a normal posterior distribution for μ

expected pattern; thus the null is highly unlikely. This is called *Probabilistic Modus Tollens* because it adds a probabilistic element ("highly likely") to the correct but deterministic logic of Modus Tollens. Consider the following example: if a person is a US citizen then it is highly unlikely she is a member of the US Congress (approximately 535/320M); the person is a member of the US Congress; therefore it is highly unlikely she is a US citizen. The key problem is that the NHST is built on the conditional probability $p(D|H_0)$, meaning the probability of seeing the data at hand given the null is true. But many interpret it as the conditional probability $p(H_0|D)$, which it is not true because the null was assumed to be true first in order to perform the test. We know from Bayes' Law that these two quantities are not equal. This also leads to common misconceptions about interpreting p-values and "stars" on regression tables. Stars are stupid because they imply that two stars (usually a p-value less than 0.01) mean that the null hypothesis is likely than when there is only one star (usually a p-value less than 0.05). Also, the dichotomization of the decision

Table 6 Counts of hate message retweets by Twitter users

Minimum	1st Quartile	2nd Quartile	3rd Quartile	Maximum
2.000	2.000	2.000	4.000	97.000

("significant" versus "non-significant") based on completely arbitrary p-value cutoffs that have no theoretical basis whatsoever is ridiculous and harmful to science. There are other problems as well, and yet it remains the most common statistical decision tool. A lively discussion about why the NHST endures in science is found in Stunt et al. (2021). The key point here for our purposes is that *Bayesian inference is completely free of all of the pathologies associated with the NHST* since all inferential statements are made with regular probability assertions such as "there is a 94% probability that the effect is positive."

8.1 Retweeting Hate Messages

As a running example for this section consider a paper by Founta et al. (2018) looking at eight months of about 80,000 abusive messages sent by Twitter users in 2017, using crowdsourcing to annotate a set of abuse-related labels, as well as other related data. For this example we take a subset of 1,960 of their collected data wherein they count the number of *retweets* of offensive, abusive, or hateful messages. There are four obvious outliers at $(102, 143, 212, 219)$ that are substantially different than the rest of the data and there is something different going on with these cases. Including such cases leads to overdispersion with Poisson modeling and there are lots of alternative model specification, which we do not discuss here for pedagogical simplicity. Our solution is to remove these four very different cases for simplicity, with the idea that interested researchers can analyze them separately. The resulting data have mean 4.76943, variance 57.22302, and sum 9329. Table 6 shows that they are also still somewhat right-skewed with a few hate messages getting many retweets.

Since these $\mathbf{y}$ are counts we employ the Gamma-Poisson conjugate model from Section 4.2:

$$p(\theta|\mathbf{y}) = (\beta + n)^{\alpha+\sum_{i=1}^{n} y_i}\Gamma\left(\alpha + \sum_{i=1}^{n} y_i\right)^{-1} \theta^{\alpha+\sum_{i=1}^{n} y_i - 1} \exp[-(\beta + n)\theta]. \quad (8.1)$$

for specified hyperpriors $\alpha = 50$ and $\beta = 10$ from the gamma prior. The prior to posterior update is shown in the first panel of Figure 9. This is an example where the data have *a lot* to say obviously. The posterior distribution is $\mathcal{G}(9379, 1966)$. The prior was constructed based on two criteria. First is the idea of mean-matching since the expected value of a set of data distributed

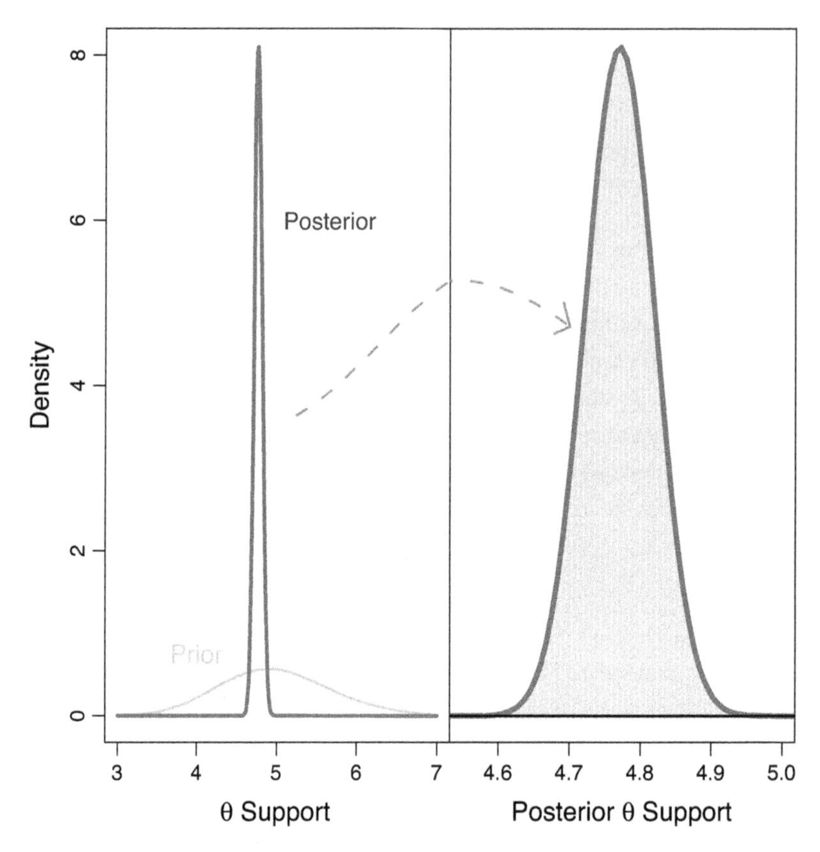

Figure 9 Prior to posterior θ intensity parameter, hate retweets

Poisson is the intensity parameter, so the prior mean was set to $50/10 \approx 4.77$. Second is the objective to specify a conservative diffuse form with the magnitude of the two hyperparameter values. It turns out, however, that this model is not very sensitive to the specification of these values because the data are large and influential in the balance between prior information and likelihood information.

The second panel of Figure 9 shows only the posterior distribution with a narrower range of the support shown at $[4.6 : 5.0]$. The shading denotes the posterior quartiles where the range between the first and second quartile (the median) is slightly narrower than the range between the second and third quartiles, indicating a little bit skewness that is not obvious from the shape of the posterior distribution alone. We can also ask questions like what is the $\pi(\theta|\mathbf{y}) > 4.75$ (or any other desired threshold)? In this case it is 0.6610584, suggesting that the model implies a strong move away from the original data mean for the Bayesian estimate of the intensity parameter with two-thirds of

Table 7 Posterior summary of θ hate retweets

Minimum	1st Quartile	2nd Quartile	3rd Quartile	Maximum
4.548	4.737	4.770	4.804	5.021

the posterior density above 4.75, which is equal to the data mean in a standard non-Bayesian MLE analysis. Since the data are large and the prior is deliberately weak, it indicates the model structure is highly influential.

8.2 Bayesian Point Estimates

Point estimates are obviously informative and are the basis of basic non-Bayesian inference. Plus, there are times when Bayesian results are intuitively conveyed by point estimates. It is important, however, to recognize that the uncertainty measure of Bayesian posteriors comes from a description of the posterior variance rather than the curvature around an MLE. Usually these are similar measures but they carry with them very different theoretical bases.

In the case of hate message retweets earlier we determined that the posterior distribution of the θ Poisson intensity parameter is given by $\mathcal{G}(9379, 1966)$, therefore:

$$\overline{\theta|\mathbf{y}} = \frac{\alpha^\dagger}{\beta^\dagger} = \frac{9379}{1966} = 4.77060 \qquad \text{Var}(\theta|\mathbf{y}) = \frac{\alpha^\dagger}{(\beta^\dagger)^2} = \frac{9379}{1966^2} = 0.00243.$$

This information is reflected in the shape of the posterior in Figure 9. Of course quantiles are a form of point estimates so we can simply report the posterior quantiles as shown in Table 7. It is very important to remember that these posterior parameter quantiles are not the same as the data quantiles given in Table 6. In the `R` and `Python` code (adjacent code boxes) these quantile values were produced by a simple form of simulation whereby 1M values are drawn randomly from the posterior distribution of θ and then analyzed as if they were regular data. With this large number of simulated values, only the minimum and the maximum will differ (slightly) on repeated trials. This idea of simulating posterior results is an extremely easy method that produces the same answer as an analytical solution, as introduced in Section 2.

R Code for Gamma Posterior Intervals

```
# SETUP
hate.retweets <- read.csv("hate.retweets.csv",
    header=FALSE)
```

```
num.retweets <- rep(NA,nrow(hate.retweets))
for (i in 1:nrow(hate.retweets))
    num.retweets[i] <- sum(!is.na(hate.retweets[i,]))
y <- num.retweets[num.retweets < 100]
n <- length(y)

# HYPERPRIOR VALUES
m <- 5; a <- 50; b <- 10

# POSTERIOR PARAMETERS
post.a <- a + sum(y)
post.b <- b + n
post.mean <- post.a/post.b
post.var <- post.a/(post.b^2)

# POSTERIOR DENSITY TO THE RIGHT OF 4.75
1-pgamma(4.75,shape=post.a,rate=post.b)

# GRAPH OF PRIOR-POSTERIOR AND POSTERIOR QUANTILES
par(oma=c(5,5,2,2), mar=c(0,0,0,0),cex.lab=1.25,
    mfrow=c(1,2))
ruler <- seq(from=3,to=7,length=500)
plot(ruler,dgamma(ruler,a,b),type="l",lwd=3,col="grey70",
    ylab="",xlim=c(3,7),ylim=c(0,8))
mtext(side=1,outer=FALSE,expression(paste(theta,
    " Support")), cex=1.5,line=3)
mtext(side=2,outer=TRUE,"Density",cex=1.5,line=3)
lines(ruler,dgamma(ruler,post.a,post.b),lwd=4,
    col="grey30")
text(4,0.75,"Prior",col="grey70",cex=1.25,adj=0.5)
text(5.72,6,"Posterior",col="grey10",cex=1.25,adj=0.5)
curve1.x <- seq(5.25,7.4,length=20)
curve1.y <- 4.50+sin(curve1.x)
lines(curve1.x,curve1.y,lwd=3,lty=2,col="red")
plot(ruler,dgamma(ruler,post.a,post.b),lwd=4,
    col="grey30",ylab="", xlab="",yaxt="n", type="l",
    xlim=c(4.55,5.0),ylim=c(0,8))
mtext(side=1,outer=FALSE,expression(paste("Posterior ",
    theta, " Support")),cex=1.5,line=3)
```

```
g.quantiles <- qgamma(c(0.25,0.5,0.75),post.a,post.b)
for (i in ruler[1:259]) segments(i,0,i,
    dgamma(i,post.a,post.b), col="grey85",lwd=5)
for (i in ruler[260:264]) segments(i,0,i,
    dgamma(i,post.a,post.b), col="grey75",lwd=5)
for (i in ruler[264:268]) segments(i,0,i,
    dgamma(i,post.a,post.b), col="grey50",lwd=5)
for (i in ruler[269:500]) segments(i,0,i,
    dgamma(i,post.a,post.b), col="grey30",lwd=5)
lines(ruler,dgamma(ruler,post.a,post.b),lwd=5,
    col="grey30")
abline(h=0,lwd=3)
curve2.x <- seq(4.40,4.70,length=20)
curve2.y <- 4.27+cos(curve1.x)
lines(curve2.x,curve2.y,lwd=3,lty=2,col="red")
segments(curve2.x[20],curve2.y[20],curve2.x[20]-0.009,
    curve2.y[20]+0.30,lwd=3,col="red")
segments(curve2.x[20],curve2.y[20],curve2.x[20]-0.022,
    curve2.y[20]-0.08,lwd=3,col="red")

# QUANTILES OF THE POSTERIOR FOR theta
n.sims <- 1000000
theta.large.sample <- rgamma(n.sims,post.a,post.b)
summary(theta.large.sample)
```

PYTHON Code for Gamma Posterior Intervals

```
import pandas as pd
import numpy as np
from scipy.stats import gamma, poisson, gaussian_kde
import matplotlib.pyplot as plt
from matplotlib.patches import ConnectionPatch

# SETUP
hate_retweets = pd.read_csv("hate.retweets.csv",
    header=None)
num_retweets = hate_retweets.count(axis=1).values
y = num_retweets[num_retweets < 100]
```

```
n = len(y)

# HYPERPRIOR VALUES
m = 5; a = 50; b = 10

# POSTERIOR PARAMETERS
post_a = a + np.sum(y)
post_b = b + n
post_mean = post_a / post_b
post_var = post_a / (post_b**2)

# POSTERIOR DENSITY TO THE RIGHT OF 4.75
p = 1 - gamma.cdf(4.75, post_a, scale=1/post_b)
print(p)

# GRAPH OF PRIOR-POSTERIOR AND POSTERIOR QUANTILES
fig, axs = plt.subplots(1, 2, figsize=(12, 6))
ruler = np.linspace(3, 7, 500)
axs[0].plot(ruler, gamma.pdf(ruler, a, scale=1/b),
    color='0.7', linewidth=3, label='Prior')
axs[0].plot(ruler, gamma.pdf(ruler, post_a,
    scale=1/post_b), color='black',
    linewidth=3, label='Posterior')
axs[0].text(4, 0.75, "Prior", color='0.7', fontsize=12)
axs[0].text(5, 6, "Posterior", color='0.3', fontsize=12)
axs[0].set_xlabel('$\Theta$  Support')
axs[0].set_ylabel('Density')
axs[0].set_ylim(-0.25, 8.25)
axs[1].plot(ruler, gamma.pdf(ruler, post_a,
    scale=1/post_b), color='black', linewidth=3)
axs[1].set_xlim(4.55, 5.0)
axs[1].set_ylim(-0.25, 8.25)
axs[1].set_xlabel('Posterior $\Theta$ Support')
axs[1].tick_params(which='both', left=False,
    labelleft=False)
axs[1].fill_between(ruler, gamma.pdf(ruler, post_a,
    scale=1/post_b), color='0.85')
con = ConnectionPatch(
   xyA=(0.5, 3.55), coordsA=axs[0].get_yaxis_transform(),
```

```
    xyB=(0.3, 5), coordsB=axs[1].get_yaxis_transform(),
    linestyle='dashed', arrowstyle="->", lw=2,
    color='red',
    connectionstyle="angle,angleA=-150,angleB=-20,rad=50")
axs[1].add_artist(con)
plt.subplots_adjust(wspace=0)
plt.show()

# QUANTILES OF THE POSTERIOR FOR theta
n_sims = 1000000
theta_large_sample = np.random.gamma(post_a, 1/post_b,
    size=n_sims)
pd.DataFrame(theta_large_sample).describe()
```

8.3 Intervals and Sets

Bayesians generally prefer to give interval estimates rather than point estimates, although some publishing outlets require point estimates by rule or culture. The Bayesian analogue of the confidence interval is the *credible interval*, which looks exactly the same except that the interpretation is completely different. The correct, but often misunderstood, definition of the $1-\alpha$ confidence interval is: an interval that over $100(1 - \alpha)\%$ of replications contains the true value of the parameter on average, as noted previously. Conversely, the definition of the Bayesian credible interval is: an interval that has a $(1 - \alpha)$ probability of containing the true value of the parameter. These are very different statements. One of the reasons that early students of statistics misinterpret the standard confidence interval is that it is built directly on the very frequentist idea of replicating the same experiment multiple times as described in Section 2, but in introductory texts it is applied to a single sample problem with an "as if" sleight-of-hand. In fact, with a confidence interval the *probability of coverage is either zero or one*, since it either covers the true θ value or it doesn't. It turns out that most of these students (and others!) who misinterpret the confidence interval are actually thinking in terms of the credible interval, but this interpretation is reserved for Bayesian inference where all unknown quantities are treated probabilistically and "confidence" has no meaning.

For the moment we will only consider unimodal posterior forms of the posterior distribution, which are very common. The Bayesian credible set is not unique since the only criterion is the $(1 - \alpha)$ coverage requirement of the posterior distribution:

$$1 - \alpha = \int_C \pi(\theta|\mathbf{y})d\theta \tag{8.2}$$

where C is a *contiguous* subset of the parameter space Θ. The coverage requirement can be met in different ways as long as the included posterior density reaches the $(1-\alpha)$ level. Most of the time in practice, it is calculated in exactly the same way as the confidence interval. For instance calculating a 95% credible interval under the Gaussian normal assumption means marching out 1.96 standard errors from the posterior mean (or mode) in either direction, just as the analogous confidence interval is usually created. However, this is most appropriate for symmetric distributions where the interpretation is easier. With an asymmetric distribution it can be confusing since there will be higher density areas left out of the interval than some equal sized intervals inside the interval. An obvious and easily interpreted alternative is to require that the x-axis areas outside the interval always have lower density above them than the areas inside the interval. More formally, a $(1-\alpha)$ *equal tail* credible interval $[L:H]$ meets the condition that:

$$\frac{\alpha}{2} = \int_{-\infty}^{L} \pi(\theta|\mathbf{y})d\theta \qquad \text{and} \qquad \frac{\alpha}{2} = \int_{H}^{\infty} \pi(\theta|\mathbf{y})d\theta, \tag{8.3}$$

which can be calculated in software with quantile functions knowing the parameters of $\pi(\theta|\mathbf{y})$. The integral limits of $(-\infty, \infty)$ earlier are given for generality. If there exist specific limits for a given posterior they are specified in these forms, such as the lower limit of zero for the gamma PDF.

We can also easily accomplish this criterion in a more general way using Monte Carlo simulation again with the 1M values drawn before. The procedure is to sort this large sample and then just pick the quantiles of interest empirically corresponding to the desired α values. This is provided in the adjacent `R` and `Python` code boxes for different α values, and the algorithm is as follows:

- generate $big.sample.size = 1\text{M}$ values (or some other large number) from the posterior distribution, and label this $\boldsymbol{\theta}_{large\ sample}$
- sort these values and save them in a new object, $\boldsymbol{\theta}_{sorted\ large\ sample}$
- for a desired α level choose the two values from this sorted vector at the positions: $\alpha/2 \times big.sample.size$, and $(1-\alpha/2) \times big.sample.size$.

Since the draws are sorted and the distribution is unimodal we are guaranteed to have the $\alpha/2$ tails identified by these two thresholds that determine where the lowest density starts and stops on either end. These values for the θ parameter in the hate message retweets model are shown in Table 8. With this tool there will be some rounding error, but with large-n datasets it is small (around $1/\sqrt{n}$).

Table 8 Credible intervals for θ in hate message retweets analysis

$1-\alpha=0.90$	$1-\alpha=0.95$	$1-\alpha=0.99$	$1-\alpha=0.999$
[4.708:4.834]	[4.690:4.852]	[4.657:4.886]	[4.620:4.925]

R Code for Gamma Posterior Credible Interval

This example continues code from Section 8.2.

```
# CREDIBLE INTERVALS FOR theta
n.sims <- 1000000
vals <- c(0.001,0.01,0.05,0.10)
sort.theta.sample <- sort(theta.large.sample)
round(sort.theta.sample[c(n.sims*vals,
    n.sims*(1-vals))],3)
```

Python Code for Gamma Posterior Credible Interval

This example continues code from Section 8.2.

```
# CREDIBLE INTERVALS FOR theta
n_sims = 1000000
vals = np.array([0.001,0.01,0.05,0.10])
s_theta_sample = np.array(sorted(theta_large_sample))
print(np.round(s_theta_sample[
    np.floor(n_sims*vals).astype(int)], 3))
print(np.round(s_theta_sample[
    np.floor(n_sims*(1-vals)).astype(int)], 3))
```

Code for Credible Interval from Drawn Data

As an illustrative numerical example, suppose we have 1M sorted numerical values sampled from a standard normal distribution and we want a 99% credible interval. Then for interval endpoints we pick the 1000000×0.005th and the 1000000×0.995th value of the sorted vector. In R:

```
n.sims <- 1000000
sort.x <- sort(rnorm(n.sims))
sort.x[n.sims*0.005]
sort.x[n.sims*0.995]
```

And in Python(with numPy):

```
n_sims = 1000000
x = np.random.normal(0, 1, n_sims)
sortx = np.sort(x)
print(sortx[int(n_sims * 0.005)])
print(sortx[int(n_sims * 0.995)])
```

What if we had a multimodal posterior form, or we had no a priori idea how many modes existed? Using either of the credible interval approaches with a multimodal prior could result in some lower density areas being included in the interval at the expense of higher density areas. There could also be non-contiguous regions of the $(1 - \alpha)$ criteria about inclusion and exclusion. To see this consider Figure 10 where there are two equal modes (although they do not have to be equal) and the $(1 - \alpha)$ range of the support has two regions. This is a difficulty for the standard credible interval calculation because it would end up containing low regions of density, particularly the one in the middle of this figure. So what we want is a tool that gives the highest $(1 - \alpha)$ density region *regardless of contiguity*. This is called the *Highest Posterior Density* (HPD) interval and it is depicted in Figure 10 by the two intervals labeled **C**. Looking at the horizontal line at k, imagine moving it up and down vertically. Moving it up increases the left-out area increasing the value of α, and moving it down decreases the left-out area decreasing α. So the way to determine the size and location of the noncontiguous regions with the highest density is to calibrate k for the desired $1-\alpha$ coverage. This approach picks the right regions of inclusion and exclusion regardless of the modality of the target distribution.

More formally, a $100(1 - \alpha)\%$ highest posterior density (HPD) interval for some θ posterior estimate is the subset of the support of the posterior distribution for the parameter θ that meets the criteria:

$$C = \{\theta : \pi(\theta|\mathbf{y}) \geq k\},$$

where k is the largest number such that:

$$1 - \alpha = \int_{\theta:\pi(\theta|\mathbf{y})\geq k} \pi(\theta|\mathbf{y})d\theta$$

The role of k in defining the area of integration is determined by the $\geq$ inequality statement, meaning that the vertical distance between the red line and PDF value in Figure 10 must be nonnegative. Since k also determines the excluded areas it is specified by α. Calculation of HPDs can be accomplished analytically but it is almost always done computationally. Consider an example where

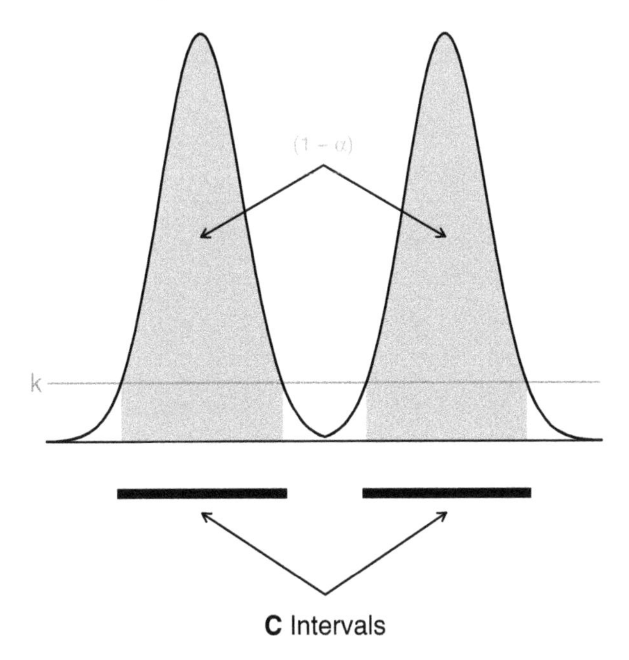

Figure 10 Highest posterior density interval for a bimodal distribution

we draw (simulate) a mixture distribution with equal numbers of $\mathcal{N}(1,1)$ and $\mathcal{N}(7,2)$ distributed values. This example is more general than it seems since we have already seen that a convenient way of calculating posterior quantities, even when we have the exact parameterization of the distribution, is using a large set of simulated draws from the distribution and calculating these quantities empirically as if they were data. Using these simulated mixture data with two modes we will implement a simple HPD algorithm that uses a density estimate to provide a crude histogram that can be summed with different values of k cutting across horizontally. Starting at the highest mode this algorithm lowers k until the α criteria is met for the HPD. This is given in the adjacent code boxes for a $1 - \alpha = 0.95$ HPD, as graphed in Figure 11.

R Code for HPD Interval from Drawn Data

```
n.sims <- 10000
mix.dat <- c(rnorm(n.sims/2,1,1), rnorm(n.sims/2,7,2) )
mix.dens <- density(mix.dat)
alpha <- 0.05
decrement <- 10000
target <- sum(mix.dens$y * mix.dens$x) * alpha
```

```
exclude <- sum(mix.dens$y * mix.dens$x)
k <- max(mix.dens$y)
while (exclude > target)  {
    k <- k - k/decrement
    exclude <- sum(mix.dens$y[mix.dens$y < k]
        * mix.dens$x[mix.dens$y < k])
    print(paste("exclude: ",exclude,"   k: ",k))
}
```

PYTHON Code for HPD Interval from Drawn Data

```
n_sims = 10000
mix_dat = np.concatenate((
    np.random.normal(1, 1, int(n_sims/2)),
    np.random.normal(7, 2, int(n_sims/2))))
mix_dens = gaussian_kde(mix_dat)
# SET n=512 TO REPLICATE density() FUNCTION IN R
dens_x = np.linspace(min(mix_dat), max(mix_dat), 512)
dens_y = mix_dens(dens_x)
alpha = 0.05
decrement = 10000
target = np.sum(dens_x * dens_y) * alpha
exclude = np.sum(dens_x * dens_y)
k = max(dens_y)
while exclude > target:
    k -= k / decrement
    mask = dens_y < k
    exclude = np.sum(dens_y[mask] * dens_x[mask])
    print(f"exclude: {exclude}   k: {k}")
```

8.4 Comparing Different Models

There are many ways to compare alternative model forms in Bayesian statistics. In this section we will concentrate on one intuitive version where alternative hyperprior values produce different models and we consider which model implies data that are closer to the observed data $\mathbf{y}$. The key principle is that since all uncertainty is described probabilistically and the end-product of a given analysis is a posterior distribution, we can use simulated draws to produce distributions for various quantities of interest. For the hate retweets data we will compare two posterior distributions for θ with the gamma-Poisson conjugate

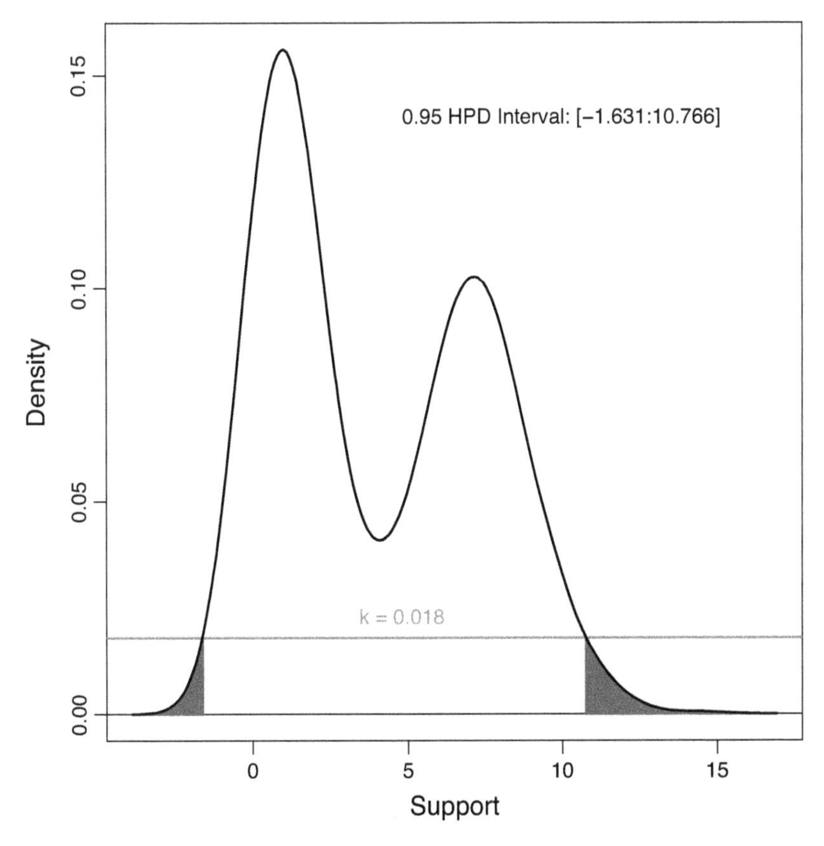

Figure 11 Highest posterior density interval from simulation

setup: one performed previously in (8.1), and another with a more diffuse prior distribution according to $\mathcal{G}(1, 1000)$. The core steps are:

- for each model to be compared draw a large number of samples from the posterior distribution for the estimated parameter (θ in this case)
- for each one of these draws plug it into the original PMF or PDF of the data and draw a simulated version of y
- compare these simulated y's with the true y's to see which model best fits the data.

With this procedure the uncertainty from the distribution of the posterior estimate is expressed in the simulated data values. Conversely, if we had just used the mean of the posterior draws this would be ignoring the distributional uncertainty that exists because the result of the Bayesian models was a posterior distribution not a point estimate.

Table 9 Summary of data and predicted data, hate retweets

	Min.	1st Q.	2nd Q.	Mean	3rd Q.	Max.
Data	2	2	2	5.1045	4	219
$m = 5, a = 50, b = 10$	2	4	5	5.2666	7	21
$m = 2, a = 1, b = 1000$	2	3	4	3.8357	5	16

Returning to the hate message retweets data we stipulate two sets of hyperpriors as shown in the first column of Table 9, where the first set of values are the ones used in the original conjugate analysis (Section 8.1) and the second set represents a more skeptical version of the same prior. These are updated according to the analysis in Section 8.1, and 100,000 draws from each posterior are produced. Using these posterior parameter values for θ we draw many data draws for the two different model setups. The results are then compared with the original data to see which model more accurately simulates data that resemble the known data. This can be done in many different ways, including graphing, but here a simple summary is given in Table 9. Notice that the original specification of the hyperparameters gives a more accurate mean summary of the predicted data. Also, both specifications struggle with predicting the larger values even after leaving out four very big outliers, which is not surprising. There is just something substantively different about the extreme cases here, and they may need to be qualitatively analyzed separately. This is obviously not an uncommon phenomenon in social science data.

R Code for Hate Retweets Model Comparison

```
m1 <- 5; a1 <- 50; b1 <- 10
m2 <- 2; a2 <- 1; b2 <- 1000
post.a1 <- a1 + sum(y); post.b1 <- b1 + n
post.a2 <- a2 + sum(y); post.b2 <- b2 + n
theta1.vals <- rgamma(100000,shape=post.a1,rate=post.b1)
theta2.vals <- rgamma(100000,shape=post.a2,rate=post.b2)
y1 <- y2 <- NULL
for (i in 1:1000)  {
   y1 <- c(y1,rpois(n,sample(theta1.vals,1,replace=TRUE)))
   y2 <- c(y2,rpois(n,sample(theta2.vals,1,replace=TRUE)))
}
y1 <- y1[y1 > 1]; y2 <- y2[y2 > 1]
rbind(summary(y),summary(y1),summary(y2))
```

PYTHON Code for Hate Retweets Model Comparison

```
m1, a1, b1 = 5, 50, 10
m2, a2, b2 = 2, 1, 1000
post_a1 = a1 + np.sum(y)
post_b1 = b1 + len(y)
post_a2 = a2 + np.sum(y)
post_b2 = b2 + len(y)
theta1_vals = gamma.rvs(a=post_a1, scale=1/post_b1,
    size=100000)
theta2_vals = gamma.rvs(a=post_a2, scale=1/post_b2,
    size=100000)
y1 = []; y2 = []
for _ in range(1000):
    y1.extend(poisson.rvs(
       mu=np.random.choice(theta1_vals, 1, replace=True),
       size=len(y)))
    y2.extend(poisson.rvs(
       mu=np.random.choice(theta2_vals, 1, replace=True),
       size=len(y)))
y1 = np.array(y1)
y2 = np.array(y2)
y1 = y1[y1 > 1]
y2 = y2[y2 > 1]
print(np.percentile(y, [0, 25, 50, 75, 100]))
print(np.percentile(y1, [0, 25, 50, 75, 100]))
print(np.percentile(y2, [0, 25, 50, 75, 100]))
```

One of the themes of this section has been that full descriptions of posterior distributions are generally more useful than point estimates in terms of describing model results. Of course point estimates can be easily produced from these full distributions if desired. This is also where the value of simulation is evident. If we have the parametric definition of the posterior distribution of interest, then generating 100,000 values or more from this distribution in `R` or `Python` is straightforward and fast with modern computers. Since all unknown quantities, including the posterior distribution, are described probabilistically in Bayesian inference, then all of this simulation work is on the probability metric and therefore intuitive. Straight likelihood inference, as described in Section 3, does not allow such direct statements such as: what is the probability that the effect is above or below some threshold of interest, what is the probability that one model fits better than another, and what is the predicted probability of future data. Furthermore, armed with many draws from a parameter's posterior distribution we have huge flexibility in calculating general summaries of interest, such as credible intervals, quantiles, and transformations. These strategies are highlighted in the detailed case study provided in the next section.

9 Case Study I: Election Polling and Bayesian Updating

In this section, we focus on a practical application of Bayesian principles and mechanics we have been discussing throughout the Element with a detailed case study. As noted, one major advantage of Bayesian statistics is its ability to incorporate prior knowledge and update with new information as it becomes available. The key tenet of Bayesian inference is to this use prior information combined with data information to produce an updated distribution for each parameter of interest as described in detail in Section 4. In this section, we will apply this principle to the context of election polling data. This process will involve using previous election results as prior belief and Bayesian updating to refine the distributional knowledge, reflecting the most current state of information.

Election polling is likely familiar to anyone who generally follows politics, in addition to political pundits and analysts. For the past several decades, since the dawn of modern survey research, polling has been an extremely important tool for gauging voter sentiment for candidates before the election date and often predicting election outcomes (Gelman, Hullman, Wlezien, & Morris, 2020; Stein et al., 2020; Stoetzer, Leemann, & Traunmueller, 2024). In the 2020 US election cycle, FiveThirtyEight tracked over 4,000 surveys by 240 pollsters, a mix of over 1,100 national and 3,000 state-level polls. For highly contested battleground states, fresh polls were released nearly daily as Election Day neared. This continually evolving information source is crucial not just for the media, tasked with deciphering campaign trends to the public, but also for political strategists deciding on campaign resource allocation and academic researchers studying political behavior dynamics and shifts. The richness of polling data comes with its own set of challenges, and users of such data should be versant in the concept of *total survey error* (Groves & Lyberg, 2010). The data is often noisy, even without considering recent issues of response rates and polling errors, so it poses significant hurdles to statistical analysis. Importantly, each poll, conducted at different points with a unique sample before the election, only provides a snapshot of voter sentiment at that particular moment.

In the face of these challenges, the Bayesian approach presents a powerful and flexible approach. It allows us to incorporate prior knowledge and adapt our predictions as new data becomes available, which is particularly advantageous when dealing with a multitude of polls released at different times. Instead of treating each poll as an isolated snapshot, the Bayesian approach allows us to iteratively update our model, absorbing the new data and refining our predictions. It also provides a natural yet systematic way to manage the balance between prior knowledge and new data, letting the data speak more loudly as it accumulates while still allowing the prior information to contribute to our understanding. Also by assigning probability distributions to our uncertainties, the Bayesian method creates a natural mechanism to account for the inherent noise and variability in polling data. Through this case study, we hope to provide a tangible demonstration of the unique strengths of analysis in managing uncertainty and making sense of complex, dynamic data.

9.1 Polling Data

Here we focus on the most recent 2020 US presidential election, which produced a wealth of polling data given the perceived closeness of the race and the dramatic politicking. The current state-of-the-art approach for using pre-election polls to gauge voter preference is *poll aggregation* that combines a large number of surveys (Lauderdale, Bailey, Blumenau, & Rivers, 2020; Madson & Hillygus, 2020), which is heavily used by media outlets such as FiveThirtyEight and The Economist. This approach can not only maximize the rich information embedded within different polls, but also mitigate the issues of variability across individual polls due to differences in methodologies, sample sizes, timing, scopes, and so on. However, the use of aggregation is not without its cautions (Isakov & Kuriwaki, 2020).

In the polling data collected by FiveThirtyEight during the 2020 election, the earliest national poll was from November 2018, during the 2018 midterm elections; the latest national and key states' polls are right before Election Day. As usual, there also tends to be a much larger number of national polls compared to individual states. They can be used to estimate the popular vote providing a broad overview of the country-wide voter but still have a gap in terms of predicting the outcomes of individual states, which is important in the electoral college system. Among states, the number of polls conducted also varies from state to state, with key battleground states such as Wisconsin and Pennsylvania having 600 to 1,000 polls and some states having only 100.

Table 10 displays the top six rows of the collection of pre-election polls from the 2020 US presidential election cycle. Each row in the dataset corresponds to a poll conducted at a certain time and state. The variables in the dataset are as follows:

Table 10 Polling data

	poll_id	pollster	state	end_date	sample_size	dem_share
1	72653	Ipsos	Arizona	2020-11-02	610	0.51
2	74946	The Political Matrix/The Listener Group	Florida	2020-11-02	966	0.479
3	72862	Trafalgar Group	Georgia	2020-11-02	1041	0.477
4	72621	PPP	Iowa	2020-11-02	871	0.505
5	72861	Trafalgar Group	Nevada	2020-11-02	1024	0.496
6	72647	Susquehanna	Pennsylvania	2020-11-02	499	0.496
7	…	…	…	…	…	…

- `poll_id`: a unique identifier for each poll
- `pollster`: the organization that conducted the poll
- `state`: the US state where the poll was conducted, where the term "National" is used when the poll was conducted across the entire United States
- `end_date`: the date when the survey ended
- `sample_size`: the number of respondents in the poll
- `dem_share`: the proportion of respondents (two-party share) who indicated they would vote for the Democratic candidate (i.e., Joe Biden).

9.2 Bayesian Setup

The objective of our analysis is to leverage multiple polls to estimate the voters' preferences of each state for the candidates before the election. To mimic the real, dynamic nature of the electoral process – as the election draws closer, fresh and more accurate polling data becomes available – we also use an updating strategy that can incorporate the new data into our estimations, continually refining our model and reflecting the evolving state of the electoral process. For the sake of pedagogical simplicity, we will not be addressing some practical issues in this area that polling practitioners are routinely concerned with, such as pollster biases, correlations between units, missing data/refusals, lack of attention, and mode of collection. Nevertheless, our analysis retains a simple, but similar, structure and can later naturally expand to more complex models and incorporate additional information. It is worth noting that the majority of those more advanced models adhere to the Bayesian framework.

To begin with, start with the basic process of using multiple polls to estimate voters' preferences for each state. In state j, the actual proportion of voters favoring the Democratic candidate is θ_j. And for a particular state poll k, there are y_{jk} voters out of the n_{jk} polled voters favoring the Democratic candidate. While we have not seen double subscripting before in this discussion, the additional complexity of notation is minor. We specify a binomial process to draw a random sample,

$$y_{jk} \sim Binomial(\theta_j, n_{jk}), \tag{9.1}$$

that was described in detail in Section 6.1 for the number of successes in a fixed number of independent individual Bernoulli trials with the same probability of success. Here, each poll is treated as a random sample from a binomial distribution, where θ_j is now the probability of success defined as a voter favoring the Democratic Party candidate in the jth state, and n_{jk} is the number of trials, meaning potential voters polled. Empirically the θ_j values are estimated from the proportion of support for the Democrat in state j starting with the binomial likelihood function, which means that we need to specify a prior for these in the Bayesian modeling process. A common choice of prior in this setting is the beta distribution since it is not only conjugate to the binomial likelihood function producing a beta form for the posterior (Table 3 in Section 5.1), it also automatically provides the appropriate bounds for the θ_j parameters. The beta PDF is given by for an arbitrary random variable X:

$$\mathcal{BE}(X|\alpha,\beta) = \frac{\Gamma(\alpha+\beta)}{\Gamma(\alpha)\Gamma(\beta)} X^{\alpha-1}(1-X)^{\beta-1}, \quad 0 < X < 1, 0 < \alpha,\beta, \tag{9.2}$$

where $\Gamma()$ denotes the Gamma function given in Section 4.2. Another virtue of this choice is that the shape of this distribution is very flexible within the support $[0:1]$ depending on the choice for α and β: it can be unimodal or bimodal, symmetric or skewed, and skewed left or right. Since we are applying this choice to each of the 50 states and Washington D.C. we are asserting $j = 1,\ldots,51$ beta priors: $\theta_j \sim \mathcal{BE}(\alpha_j,\beta_j)$, with different versions of α_j and β_j. To inform these prior choices for our analysis of the 2020 election we turn to the 2016 presidential election to provide an informed prior baseline.

For each state j and poll k combination we have observed y_{jk} for the Democratic candidate support and n_{jk} for the number of respondents, in that state/poll combination. This a basic example of a Bayesian hierarchical model, which are very powerful ways to analyze data from different sources and at different levels of aggregation. In the following empirical example we will confine ourselves to a single state (Georgia) or individual states separately for the pedagogical purpose of simplicity but more elaborate Bayesian models can easily incorporate this hierarchy, and this will be done later. This decision means that the j notation is superfluous but we leave it in the following expressions as a reminder of Bayesian modeling flexibility. The likelihood function for this setup thus sums across the K polls selected for any given model available for use in a state j:

$$L(\theta_j|y_{jk}, n_{jk}) = \prod_{k=1}^{K} \binom{n_{jk}}{y_{jk}} \theta_j^{y_{jk}} (1-\theta_j)^{n_{jk}-y_{jk}}. \tag{9.3}$$

The double subscripting jk is a reminder of the state j and poll k crossing.

The posterior distribution for a single state j is given proportionately by the beta prior times the binomial likelihood function:

$$\pi(\theta_j|y_{jk}, n_{jk}, \alpha_j, \beta_j) = \frac{\Gamma(\alpha_j + \beta_j)}{\Gamma(\alpha_j)\Gamma(\beta_j)}\theta_j^{\alpha_j-1}(1-\theta_j)^{\beta_j-1} \times \prod_{k=1}^{K}\binom{n_{jk}}{y_{jk}}\theta_j^{y_{jk}}(1-\theta_j)^{n_{jk}-y_{jk}}$$

[use proportionality to dispose of constants]

$$\propto \theta_j^{\alpha_j-1}(1-\theta_j)^{\beta_j-1}\prod_{k=1}^{K}\theta_j^{y_{jk}}(1-\theta_j)^{n_{jk}-y_{jk}}$$

[move the products into the exponents as sums]

$$= \theta_j^{\alpha_j-1}(1-\theta_j)^{\beta_j-1}\theta_j^{\sum_{k=1}^{K} y_{jk}}(1-\theta_j)^{\sum_{k=1}^{K}(n_{jk}-y_{jk})}$$

[collect terms]

$$= \theta_j^{\alpha_j-1+\sum_{k=1}^{K} y_{jk}}(1-\theta_j)^{\beta_j-1+\sum_{k=1}^{K}(n_{jk}-y_{jk})}.$$

To see the resulting distributional form we can define:

$$\text{something} = \alpha_j + \sum_{k=1}^{K} y_{jk}$$

$$\text{something else} = \beta_j + \sum_{k=1}^{K}(n_{jk} - y_{jk})$$

so that the last line in the posterior calculation earlier is:

$$\pi(\theta_j|y_{jk}, n_{jk}, \alpha_j, \beta_j) \propto \theta_j^{\text{something}-1}(1-\theta_j)^{\text{something else}-1}.$$

Now it is straightforward to see from the definition in (9.2) that the derived posterior distribution is the kernel of another beta distribution given by $\mathcal{BE}(\alpha_j + \sum_{k=1}^{K} y_{jk},\ \beta_j + \sum_{k=1}^{K}(n_{jk} - y_{jk}))$. Notice again the obvious compromise here between prior information (α_j, β_j) and data information $\left(\sum_{k=1}^{K} y_{jk},\ \sum_{k=1}^{K}(n_{jk} - y_{jk})\right)$, which is always present in Bayesian inference. Now we have a complete recipe for analyzing and summarizing the posterior distribution of θ_j given that we have $(y_{jk}, n_{jk}, \{\alpha_j, \beta_j\})$, and no more analytical work is required.

With the posterior distribution fully determined, we now can think about how prior belief $(\{\alpha_j,\ \beta_j\})$ and data information $(\mathbf{y}_{jk},\ \mathbf{n}_{jk})$ collectively affect the posterior distribution. The mean of a variable X distributed $\mathcal{BE}(\alpha, \beta)$ is $E[X] = \alpha/(\alpha+\beta)$, and the variance is $\text{Var}[X] = (\alpha\beta)/((\alpha+\beta)^2(\alpha+\beta+1))$. Another distributional summary of dispersion is the "concentration," denoted by $\kappa = \alpha + \beta$, and is the extent to which density is concentrated near the mean. As demonstrated in Figure 12, a larger κ makes the beta distribution narrower and more concentrated. Consequently, the selection of $\kappa = \alpha + \beta$ for the prior

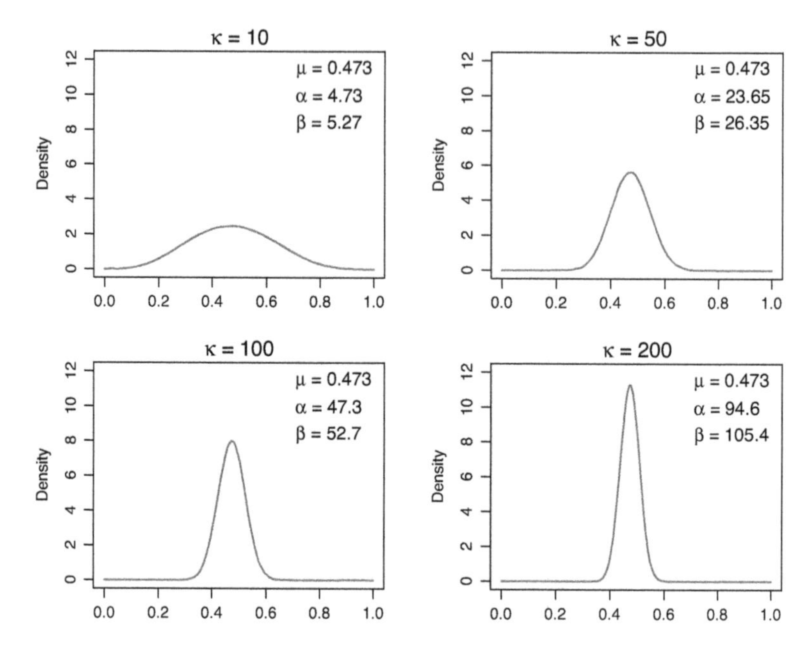

Figure 12 Beta distributions with different concentrations

has further implications. It sets the tone for how "assertive" or "persuasive" our prior beliefs appear when confronted with new data. Furthermore, for posterior beta distribution, each new data point – a tally of support for Democratic or Republican candidates – incrementally augments $\sum_{k=1}^{K} y_{jk}$ and $\sum_{k=1}^{K}(n_{jk} - y_{jk})$). Given the substantial sample size often inherent in surveys – running into hundreds or thousands – the influx of new data information can dramatically sway our prior beliefs and result in a rapid reduction in uncertainty. To put it in perspective, we probably would not think another 1000 respondents' opinions are equivalent to another 1000 coin flips in terms of reducing uncertainty.

In this light, instead of representing raw count, it makes more sense to consider the parameters of the beta distribution as "weights" for success's and failure's contributions to the distribution. Therefore, we can scale down n_{jk} and correspondingly y_{jk} to reflect our uncertainty. In the following analysis, we impose a $w_0 = 0.1$ base weight for each poll, which translates to each poll carrying just a tenth of its original weight in the posterior distribution – a subtle adjustment that enables us to accommodate our degree of uncertainty. Going back to the prior distribution, we are now in a position to fine-tune the degree of influence we wish our prior beliefs to exert. A smaller $\kappa = \alpha + \beta$ implies a more malleable prior, easily swayed by new data; and to make our prior beliefs have stronger input, we would specify large $\kappa = \alpha + \beta$. For this example, we choose a $\kappa = 500$, which is equivalent to a survey with an unweighted sample size of

5000. This prior would exert a somewhat notable influence on states with few polls but will fade in on heavily polled states. This is precisely the reason why certain states have only few polls – for deep red or blue states, there is not much additional information polls can provide compared to historical election data.

Next, to bring the model closer to reality and reflect the dynamics of polling, we also impose a temporal weight $w(t)$ on the posterior variables. It is reasonable to think that polls conducted several months or even more than a year prior to the election are weak predictors of the final results. This temporal weighting allows us to prioritize and emphasize more recent polls, which are generally considered to provide a better reflection of the current public opinion. We use a long memory weight function (Christensen and Florence, 2008):

$$w_t = \begin{cases} 1 - \frac{t}{70}, t \leq 56 \\ 0.2, t > 56 \end{cases}$$

With these weights in place, our posterior distribution now becomes

$$\pi(\theta_j | y_{jk}, n_{jk}, \alpha_j, \beta_j) = \theta_j^{\alpha_j - 1 + \sum_{k=1}^{K} w_0 w_t y_{jk}} (1 - \theta_j)^{\beta_j - 1 + \sum_{k=1}^{K} w_0 w_t (n_{jk} - y_{jk})}$$

where $w_0 w_t y_{jk}$ and $w_0 w_t (n_{jk} - y_{jk})$ are simply weighted versions of the original binomial parameters. While n and y in the binomial distribution technically require integers, this weighted version is justifiable in the sense that we care more about the data proportions (y/n, i.e., the proportion of voters supporting Biden) instead of the raw count. And this will be clearer in the simulation process later.

9.3 Implementation in Software

The strategy is to use simulation again by generating a large number of samples of θ_j from this beta distribution computationally. For the 2020 US presidential election a state of high interest is Georgia since the Democrat beat the Republican by merely 12,670 votes and a margin of 0.239% (according to an audit required by state law). So we use $(\theta_{\text{GA}}, y_{\text{GA},k}, n_{\text{GA},k})$ for the empirical example with $\alpha = 95$, $\beta = 105$ ($\kappa = \alpha + \beta = 200$) prior parameters. We chose these values using the 2016 election results where the then Democratic candidate Hillary Clinton received 47.3% of the two-party vote share ($95/200 = 0.473$). It is important to note that the choice of $\kappa = 200$ is somewhat arbitrary but represents a relatively informative prior since it would result in a narrower beta distribution. The adjacent code boxes show the calculation steps of the posterior distribution for the Biden (Democrat) vote share with poll aggregation.

R Code for Calculating Georgia Posterior Distributions

```
# POLLING DATA
polls <- read.csv("polls2020.csv")
polls$end_date <- as.Date(polls$end_date)
ga <- subset(polls, state=="Georgia")

# KAPPA, BASE WEIGHT, etc.
kappa <- 500; w0 <- 0.1
ruler <- seq(0,1,length=5000)
days <- as.numeric(as.Date("2020-11-02")-ga$end_date)

# PRIOR
alpha <- 0.473*kappa; beta <- kappa-alpha
prior <- dbeta(ruler,alpha,beta)

# LONG MEMORY TEMPORAL WEIGHT
wt <- ifelse(days > 56, 0.2, 1-days/70)

# COMPUTE POSTERIOR
sum_n <- sum(ga$sample_size*w0*wt)
sum_y <- sum(ga$dem_share*ga$sample_size*w0*wt)
post_alpha <- alpha+sum_y
post_beta <- sum_n-sum_y+beta
posterior <- dbeta(ruler,post_alpha,post_beta)
```

PYTHON Code for Calculating Georgia Posterior

```
# POLLING DATA
polls = pd.read_csv("polls2020.csv")
polls['end_date'] = pd.to_datetime(polls['end_date'])
ga=polls[polls["state"]=="Georgia"]

# DEFINE KAPPA AND BASE WEIGHT
kappa,w0=500,0.1
ruler=np.linspace(0,1,5000)
days=(pd.to_datetime("2020-11-02")
    - ga['end_date']).dt.days
```

```
# PRIOR
alpha,beta=0.473*kappa,kappa-0.473*kappa
prior=stats.beta.pdf(ruler,alpha,beta)

# LONG MEMORY TEMPORAL WEIGHT
wt=np.where(days>56,0.2,1-days/70)

# COMPUTE POSTERIOR
sum_n=(ga['sample_size']*w0*wt).sum()
sum_y=(ga['dem_share']*ga['sample_size']*w0*wt).sum()
post_alpha=alpha+sum_y
post_beta=sum_n-sum_y+beta
posterior=stats.beta.pdf(ruler,post_alpha,post_beta)
```

We can also calculate the point estimate and highest posterior density intervals for Biden's two-party vote share. For a beta distribution, the mean (expected value) is calculated as $E[X] = \alpha/(\alpha + \beta)$ and thus does not require simulation. For this posterior mean is 0.514, which reflects the small margin of victory. The HPDs for four levels of $1 - \alpha$ are given in Table 11. It is clear that the data from the aggregated polls have a lot to say in the prior versus data trade-off since the resulting posterior intervals at any of these four levels are narrow. It is also worth noting that this beta distribution is also slightly right-skewed.

R Code for Estimates of Biden's Support

```
# POINT ESTIMATE
post_mean <- post_alpha/(post_alpha+post_beta)

# CREDIBLE INTERVALS
n_sims <- 1000000
post_samples <- rbeta(n_sims, post_alpha,post_beta)
sorted_samples <- sort(post_samples)
vals <- c(0.001,0.01,0.05,0.10)
round(sorted_samples[c(n_sims*vals, n_sims*(1-vals))],4)

# USE HDInterval PACKAGE
library(HDInterval)
hdi(post_samples,credMass=0.95)
sapply(1-vals,
       function(vals) hdi(post_samples,credMass = vals))
```

PYTHON Code for Estimates of Biden's Support

```
# POINT ESTIMATE
post_mean = post_alpha / (post_alpha + post_beta)
print(post_mean)

# CREDIBLE INTERVALS
n_sims = 1000000
post_samples = stats.beta.rvs(post_alpha, post_beta,
                              size=n_sims)
sorted_samples = np.sort(post_samples)
vals = np.array([0.001, 0.01, 0.05, 0.1])
lower = [np.round(sorted_samples[int(n_sims*val)],3)
         for val in vals]
upper = [np.round(sorted_samples[int(n_sims*(1-val))],3)
         for val in vals]
print(list(zip(lower, upper)))

# USE az.hdi() TO CALCULATE THE HDI
hdi = az.hdi(post_samples, hdi_prob=0.95)
print(hdi)

## ITERATE THROUGH EACH VALUE
ci = [az.hdi(post_samples, hdi_prob=val)
         for val in 1 - vals]
print(ci)

print("99.9%: ",
      f"[{round(ci[0][0],4)},{round(ci[0][1],4)}]")
print("99%: ",
      f"[{round(ci[1][0],4)},{round(ci[1][1],4)}]")
print("95%: ",
      f"[{round(ci[2][0],4)},{round(ci[2][1],4)}]")
print("90%: ",
      f"[{round(ci[3][0],4)},{round(ci[3][1],4)}]")
```

9.4 The Dynamics of Election and Posterior Update

Our previous analysis leveraged all available polls leading up to Election Day, aiming to estimate the overall support for Biden in Georgia. However, both

Table 11 Credible intervals for estimated Biden support in Georgia

$1-\alpha=0.90$	$1-\alpha=0.95$	$1-\alpha=0.99$	$1-\alpha=0.999$
[0.5101:0.5207]	[0.509:0.5217]	[0.507:0.5236]	[0.5048:0.5261]

the election and polling are inherently dynamic processes. As the election progresses, more data becomes available and potentially offers more accurate insights into the state of the race as they are fielded closer to the final election. To show this evolving nature of the election and polling, we can take advantage of the Bayesian approach that reflects a natural balance between prior and data and adapt our analysis to function more dynamically.

For any given point in time, we can use all polls conducted up to that date to estimate voter preferences. As we move closer to the election, we include more polls and more recent polls in our model. This means that as new polls are conducted, our estimate of voter preferences is continually updated to reflect the most current data. The adjacent code boxes implement this process in both `R` and `Python`. We begin by establishing a sequence of months over which we will examine the polls. The loop iterates over each month in the sequence. For each month, it selects all polls conducted in Georgia up to that month.

`R` Code for Dynamic Estimation of Voter Preferences

```
months <- seq(1,11,1)
ruler <- seq(0,1,length=5000)

# PLACEHOLDERS TO STORE VALUES
posteriors <- matrix(NA, nrow=length(months),ncol=5000)
post_est <- rep(NA, length(months))
post_ci <-  matrix(NA, nrow=length(months),ncol=2)

# PRIOR
alpha <- 0.473*kappa; beta <- kappa-alpha
priors <- dbeta(ruler,alpha,beta)

# ITERATE THROUGH EACH MONTH
for (m in 1:length(months)){
  today <- as.Date(paste0("2020-",months[m],"-03"))
  ga <- polls[polls$state=="Georgia" &
        polls$end_date <= today,]
```

```
  days <- as.numeric(max(ga$end_date)-ga$end_date)
  # WEIGHTS
  w0 <- 0.1
  wt <- ifelse(days > 56, 0.2, 1-days/70)
  # POSTERIOR
  sum_n <- sum(ga$sample_size*w0*wt)
  sum_y <- sum(ga$dem_share*ga$sample_size*w0*wt)
  post_alpha <- alpha+sum_y
  post_beta <- sum_n-sum_y+beta
  posteriors[m,] <- dbeta(ruler,post_alpha,post_beta)
  # ESTIMATES AND CREDIBLE INTEVALS
  post_est[m] <- post_alpha / (post_alpha+post_beta)
  post_ci[m,] <- hdi(rbeta(n_sims,post_alpha,post_beta),
        credMass=0.95)
}
```

PYTHON Code for Dynamic Estimation of Voter Preferences

```
months = np.arange(1, 12, 1)
ruler = np.linspace(0, 1, 5000)

# PLACEHOLDERS TO STORE VALUES
posteriors = np.empty((len(months), 5000))
post_est = np.empty(len(months))
post_ci = np.empty((len(months), 2))

# PRIOR
alpha = 0.473 * kappa
beta = kappa - alpha
priors = stats.beta.pdf(ruler, alpha, beta)

# ITERATE THROUGH EACH MONTH
for m in range(len(months)):
  today = pd.to_datetime(f"2020-{months[m]}-03")
  ga = polls[(polls['state']=='Georgia') &
        (polls['end_date']<=today)]
  days = (ga['end_date'].max()-ga['end_date']).dt.days
  # WEIGHTS
  w0 = 0.1
```

```
wt = np.where(days>56,0.2,1-days/70)
sum_n = (ga['sample_size']*w0*wt).sum()
sum_y = (ga['dem_share']*ga['sample_size']*w0*wt).sum()
post_alpha = alpha+sum_y
post_beta = sum_n-sum_y+beta
posteriors[m, :] = stats.beta.pdf(ruler,
      post_alpha, post_beta)
post_est[m] = post_alpha/(post_alpha+post_beta)
post_ci[m, :] = az.hdi(stats.beta.rvs(post_alpha,
      post_beta, size=n_sims), hdi_prob=0.95)
```

Figure 13 illustrates the evolution of posterior distributions of Biden's support, demonstrating how the accumulation of polling data over time refines our estimations. In the early stages, ten months prior to the election, when there were only few polls available, the posterior distribution is more spread, and our inference is largely influenced by the prior at this stage. As the election nears and more polls become available, the posterior distribution gradually narrows showing increased precision. The influence of the prior diminishes as the freshly obtained polling data provides updated insights into the evolving state of the race. In other words, the posterior distribution more accurately reflects the current state of the race. This is also reflected in Figure 14, which we frequently see in the election coverage. As Election Day approaches, the credible intervals of the estimates become increasingly narrower with more polls and more updated information.

9.5 Simulation and Election Forecasting

Based on the example of Georgia, we can now expand our analysis to all US states. Given that the electoral college system in the United States is state-based, with each state contributing a certain number of electoral votes toward the total, we can simply replicate the previous process for each state, calculating a posterior distribution for each state during different periods over the election cycle. Eventually, we can obtain an estimate of the vote share (θ_j) for all 51 states and Washington D.C. For each state, we use their respective previous election results as prior with a fixed $\kappa = 500$. We then establish a sequence of dates from March to November 2020, over which we will examine the polls. Once we obtain the estimated support for Biden per state, we can use these estimates as probabilities in binomial distributions to simulate the electoral support. Specifically, for each θ_j estimate, we draw 50,000 values from the posterior distribution as probability and plug them in binomial distribution

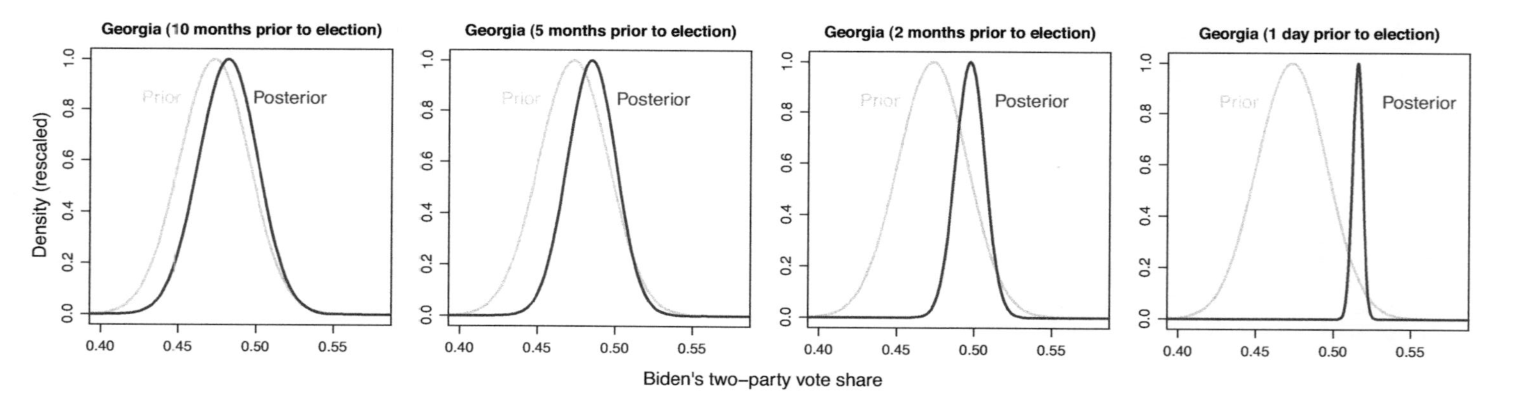

Figure 13 Updating posterior distributions with more polling data

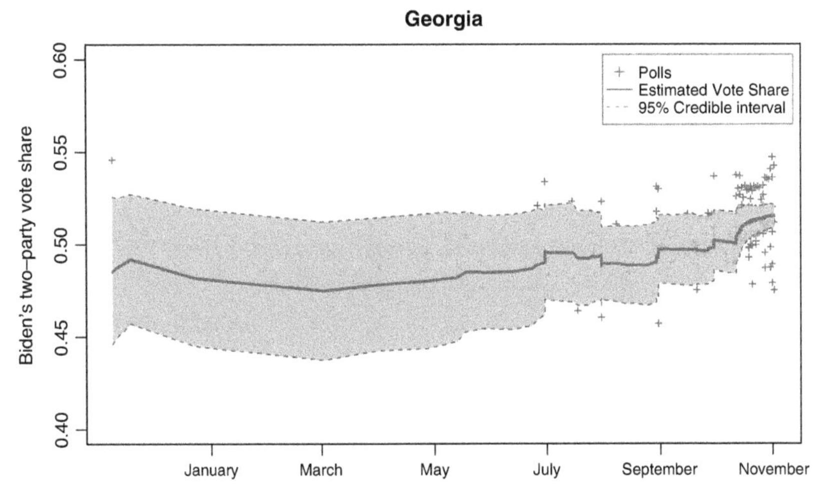

Figure 14 95% credible intervals over time

with a $N = 5000$, which is equivalent to 5,000 Bernoulli trials (i.e., voters in our case). As we are concerned more about the proportions of outcomes rather than the absolute number, the number of N does not matter much as long as they are large enough. Then, if over 50% of the voters support Biden in simulation, we will assume that he wins the state and its electoral college votes (without considering the minor exceptions of Maine and Nebraska, although it is not difficult to simulate the split electoral votes with additional data of the polls and the previous election results at the congressional district level). We will then sum up the electoral votes of the states won by the Democratic candidates to get the total number of electoral votes. This essentially creates a process from raw support to predicted electoral college outcomes.

The adjacent code boxes implement this process in both `R` and `Python`. We first load a dataset that contains the previous election results and electoral college votes for each state. The two loops iterate over each state and each specified date. For each date, it selects all polls conducted up to that date. Finally, we simulate random beta-distributed values based on posterior distribution and use the simulated value as probability to generate binomial distributions. Then, using these results we can assign the electoral college votes based on whether Biden receives 50% and more support or not. Figure 15 presents the forecasting results from March to November 2020.

R Code for Simulating Electoral Outcomes of the U.S.

```
# 2016 ELECTION AND ELECTORAL COLLEGE DATA
election2016 <- read.csv("data/election2016.csv")
```

```
states <- unique(election2016$state)
n_sims <- 50000; n_state <- length(states)
# DATES
months <- seq(4,11,1)
dates <- as.Date(c(paste0("2020-",months,"-03"),
    paste0("2020-",months[-length(months)],"-10"),
    paste0("2020-",months[-length(months)],"-17"),
    paste0("2020-",months[-length(months)],"-24")))
dates <- dates[order(dates)]; n_date <- length(dates)
# KAPPA & BASE WEIGHT
kappa <- 500; w0 <- 0.1

sim_res <- array(data=NA,dim=c(n_state,n_sims,n_date))
for (i in 1:n_state){
  cat(states[i], "\n")
  state_dat <- polls[polls$state==states[i],]
  state_dat <- state_dat[order(state_dat$end_date),]
  alpha <- election2016[election2016$state==
                          states[i],]$dem_share*kappa
  beta <- kappa - alpha
  for (d in 1:length(dates)){
    sub_dat <- state_dat[state_dat$end_date < dates[d],]
    if (nrow(sub_dat) > 0) {
      days <- as.numeric(max(sub_dat$end_date)-
                           sub_dat$end_date)
      wt <- ifelse(days > 56, 0.2, 1-days/70)
      sum_n <- sum(sub_dat$sample_size*w0*wt)
      sum_y <- sum(sub_dat$dem_share*
                     sub_dat$sample_size*w0*wt)
      post_alpha <- alpha+sum_y
      post_beta <- sum_n-sum_y+beta
    } else {
      # FOR STATES WITH NO POLLS YET, USE PRIOR
      post_alpha <- alpha
      post_beta <- beta
    }
    # SIMULATE THETA FROM POSTERIOR
    pis <- rbeta(n_sims, post_alpha, post_beta)
    # DRAW SIMULATED BINOMIAL OUTCOMES
```

```
    outcomes <- rbinom(length(pis), 2000, pis)/2000 > 0.5
    # CALCULATE ELECTORAL COLLEGE VOTES
    evs <- ifelse(outcomes,
      election2016[election2016$state==states[i],]$ev,
      0)
    sim_res[i,,d] <- evs
  }
}
# CALCULATE BIDEN'S CHANCE OF WINNING
ev_sum <- apply(sim_res, c(2,3), sum)
biden_win <- apply(ev_sum, 2, function(x) sum(x>=270))/
             n_sims
```

PYTHON Code for Simulating Electoral Outcomes of the U.S.

```
# 2016 ELECTION AND ELECTORAL COLLEGE DATA
election2016 = pd.read_csv(
    "election2016.csv")
states = election2016['state'].unique()
n_sims, n_state = 10000, len(states)
# DATES
months = range(4, 12)
dates = pd.to_datetime(
    np.concatenate([
        [f"2020-{m}-03" for m in months],
        [f"2020-{m}-10" for m in months[:-1]],
        [f"2020-{m}-17" for m in months[:-1]],
        [f"2020-{m}-24" for m in months[:-1]],
    ])
).sort_values()
n_date = len(dates)
# DEFINE KAPPA
kappa, w0 = 500, 0.1
# SIMULATION
sim_res = np.empty((n_state, n_sims, n_date))
# ITERATE THROUGH DATES AND STATES
for i, state in enumerate(states):
    s_dat = polls[polls['state'] == state]
    s_dat = s_dat.sort_values('end_date')
```

```
        d_share = election2016[election2016['state']==state]
        alpha = d_share['dem_share'].item()*kappa
        beta = kappa-alpha
        for d, date in enumerate(dates):
            sub_dat = s_dat[s_dat['end_date']<dates[d]]
            if not sub_dat.empty:
                days = (sub_dat['end_date'].max() -
                        sub_dat['end_date']).dt.days
                w0 = 0.1
                wt = np.where(days >56,0.2,1-days/70)
                sum_n = (sub_dat['sample_size'] *
                    w0*wt).sum()
                sum_y = (sub_dat['dem_share'] *
                    sub_dat['sample_size'] * w0*wt).sum()
                p_a, p_b = alpha+sum_y, sum_n-sum_y+beta
            else:
            # FOR STATES WITH NO POLLS YET, USE PRIOR
                p_a, p_b = alpha, beta
            # SIMULATE THETA FROM POSTERIOR
            pis = stats.beta.rvs(p_a,p_b,size=n_sims)
            # DRAW SIMULATED OUTCOMES FROM THE BINOMIAL
            outcomes = stats.binom.rvs(n=5000,p=pis,
                size=n_sims)/5000>0.5
            # CALCULATE ELECTORAL COLLEGE VOTES
            evs = np.where(outcomes,
                d_share['ev'].item(), 0)
            sim_res[i, :, d] = evs
    ev_sum = sim_res.sum(axis=0)
    # CALCULATE BIDEN'S CHANCE OF WINNING
    biden_win = np.apply_along_axis(
        lambda x: np.bincount(x>=270, minlength=2),
            0, ev_sum)/n_sims
```

With this case study, we demonstrate that the Bayesian framework provides a flexible yet realistic way of accounting for the diverse and dynamic nature of elections and polling. This Bayesian approach enables the combination of prior beliefs from previous elections with dynamically observed new data in a coherent and principled manner. The prior distributions based on previous election results reflect our initial beliefs about the parameters of interest (i.e., the proportion of voters favoring a candidate in each state), which is particularly

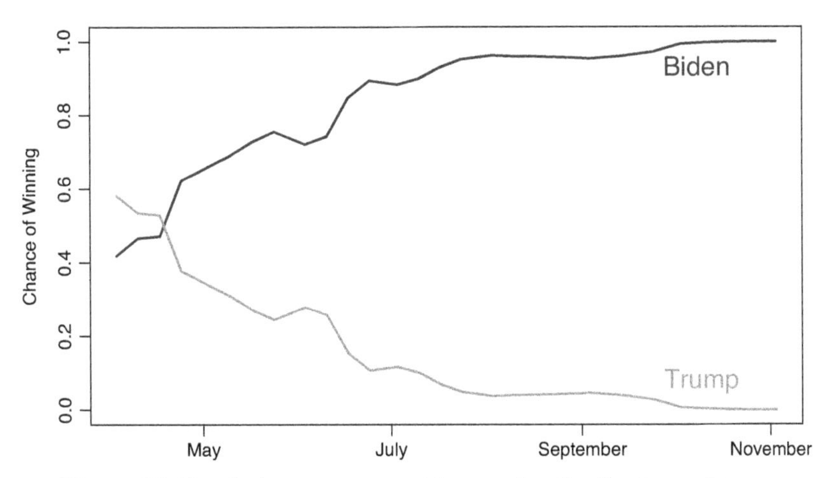

Figure 15 Simulating outcomes with posterior distributions of states

important when there is not much data available for the state. We then update our beliefs as more data become available to obtain posterior distributions. One of the notable advantages of Bayesian analysis is its ability to provide probabilistic estimates of the parameters of interest. This is particularly demonstrated in our final simulation analysis, where we obtain the posterior distributions instead of one single estimate and use a range of plausible values to simulate the electoral outcomes. This seamless integration provides a simple, yet comprehensive, analysis of the polling results.

This case study, using aggregated polling data, can be seen as a simpler version of the more advanced polling and election forecasting techniques. We use the beta distribution as the prior distribution and its conjugacy with binomial distribution to produce a beta form for the posterior. But it can be demonstrated using the Central Limit Theorem, as y and $n-y$ become large with fixed α and β, $E(\theta|y) \approx y/n$ and $var(\theta|y) \approx \frac{1}{n}\frac{y}{n}\left(1-\frac{y}{n}\right)$ approaches zero. This often justifies the use of normal distribution to approximate the posterior distribution. In fact, we can also transform the binomial parameter θ to a logit scale, making it more appropriate for a normal approximation (Gelman et al., 2015). This essentially expands our analysis to the state-of-the-art election forecasting technique that uses both polling results and "fundamentals" (i.e., structural factors that influence voter decisions) and takes into account various correlations, where Bayesian approach also plays a crucial role.

To conclude, this section highlights the value of Bayesian analysis by leveraging prior beliefs, updating them with observed data, and quantifying uncertainty. It underscores the flexibility, coherence, and probabilistic nature of Bayesian methods. These features make Bayesian analysis a powerful tool in social science studies. The concepts explored in this section and throughout the discussion lay the groundwork for more advanced techniques and highlight the value of the current Element in guiding readers through the intricacies of Bayesian analysis in the context of elections and beyond.

References

Bartels, L. M. (1996). Pooling disparate observations. *American Journal of Political Science*, *40*(3), 905–942.

Bayes, T. (1763). Lii. An essay towards solving a problem in the doctrine of chances. By the late Rev. Mr. Bayes, frs communicated by Mr. Price, in a letter to John Canton, amfr s. *Philosophical Transactions of the Royal Society of London*, (53), 370–418.

Berk, R. A., Western, B., & Weiss, R. E. (1995). Statistical inference for apparent populations. *Sociological Methodology*, 421–458.

Birnbaum, A. (1962). On the foundations of statistical inference. *Journal of the American Statistical Association*, *57*(298), 269–306.

Canes-Wrone, B., Brady, D. W., & Cogan, J. F. (2002). Out of step, out of office: Electoral accountability and house members' voting. *American Political Science Review*, *96*(1), 127–140.

Christensen,W. F., & Florence, L.W. (2008). Predicting presidential and other multistage election outcomes using state-level pre-election polls. *The American Statistician*, *62*(1), 1–10.

Copas, J. (1969). Compound decisions and empirical bayes. *Journal of the Royal Statistical Society: Series B (Methodological)*, *31*(3), 397–417.

Dale, A. I. (2012). *A history of inverse probability: From Thomas Bayes to Karl Pearson*. Springer Science & Business Media.

Diaconis, P., & Freedman, D. (1986). On the consistency of bayes estimates. *The Annals of Statistics*, *14*(1), 1–26.

Fisher, R. A. (1922). On the mathematical foundations of theoretical statistics. *Philosophical Transactions of the Royal Society of London. Series A, Containing Papers of a Mathematical or Physical Character*, *222*(594–604), 309–368.

Fisher, R. A. (1925). Theory of Statistical Estimation. *Mathematical Proceedings of the Cambridge Philosophical Society*, *22*(5), 700–725.

Founta, A.-M., Djouvas, C., Chatzakou, D. et al. (2018). Large scale crowdsourcing and characterization of Twitter abusive behavior. In *11th international conference on web and social media, ICWSM 2018*.

Gelman, A., Carlin, J. B., Stern, H. S., Dunson, D. B., Vehtari, A., & Rubin, D. B. (2015). *Bayesian data analysis* (3rd ed.). CRC Press.

Gelman, A., Hullman, J., Wlezien, C., & Morris, G. E. (2020). Information, incentives, and goals in election forecasts. *Judgment and Decision Making, 15*(5), 863–880.

Gill, J. (1999). The insignificance of null hypothesis significance testing. *Political Research Quarterly, 52*(3), 647–674.

Gill, J. (2014). *Bayesian methods: A social and behavioral sciences approach* (Vol. 20). CRC Press.

Gill, J., & Freeman, J. R. (2013). Dynamic elicited priors for updating covert networks. *Network Science, 1*(1), 68–94.

Gill, J., & Torres, M. (2019). *Generalized linear models: A unified approach* (Vol. 134). Sage.

Gill, J., & Walker, L. D. (2005). Elicited priors for Bayesian model specifications in political science research. *The Journal of Politics, 67*(3), 841–872.

Groves, R. M., & Lyberg, L. (2010). Total survey error: Past, present, and future. *Public Opinion Quarterly, 74*(5), 849–879.

Isakov, M., & Kuriwaki, S. (2020). Towards principled unskewing: Viewing 2020 election polls through a corrective lens from 2016. *Harvard Data Science Review, 2*(4).

Kolmogorov, A. N. (1933). Grundbegriffe der wahrscheinlichkeitreichnung. *Ergebnisse der Mathematik.*

Lauderdale, B. E., Bailey, D., Blumenau, J., & Rivers, D. (2020). Model-based pre-election polling for national and sub-national outcomes in the US and UK. *International Journal of Forecasting, 36*(2), 399–413.

Leamer, E. E. (1972). A class of informative priors and distributed lag analysis. *Econometrica: Journal of the Econometric Society, 40*(6), 1059–1081.

Madson, G. J., & Hillygus, D. S. (2020). All the best polls agree with me: Bias in evaluations of political polling. *Political Behavior, 42*(4), 1055–1072.

Stein, R. M., Mann, C., Stewart III, C. et al. (2020). Waiting to vote in the 2016 presidential election: Evidence fromamulti-county study. *Political Research Quarterly, 73*(2), 439–453.

Stigler, S.M. (1982). Thomas Bayes's Bayesian inference. *Journal of the Royal Statistical Society: Series A (General), 145*(2), 250–258.

Stigler, S. M. (1983). Who discovered Bayes's theorem? *The American Statistician, 37*(4a), 290–296.

Stoetzer, L. F., Leemann, L., & Traunmueller, R. (2024). Learning from polls during electoral campaigns. *Political Behavior, 46*(1), 543–564.

Stunt, J., van Grootel, L., Bouter, L., Trafimow, D., Hoekstra, T., & de Boer, M. (2021). Why we habitually engage in null-hypothesis significance testing: A qualitative study. *Plos One, 16*(10), e0258330.

Wagner, K., & Gill, J. (2005). Bayesian inference in public administration research: Substantive differences from somewhat different assumptions. *International Journal of Public Administration*, *28*(1–2), 5–35.

Zellner, A. (1996). *An introduction to Bayesian inference in econometrics*. Wiley.

Acknowledgments

We acknowledge the support received from the Center for Data Science at American University and the Massive Data Institute at Georgetown University. We also thank the editors of the series and two anonymous reviewers for comments and suggestions.

Data Availability Statement

All code and data accompanying this Element are stored in the GitHub repository: https://github.com/jgill22/Bayesian.Social.Science.Statistics, and can also be run interactively via Code Ocean: https://codeocean.com/capsule/8772484. And we confirm we want to keep the Code Ocean capsule where we mention it in the text, i.e. after the end of Chapter/Section 1.

About the Series

The Elements Series Quantitative and Computational Methods for the Social Sciences contains short introductions and hands-on tutorials to innovative methodologies. These are often so new that they have no textbook treatment or no detailed treatment on how the method is used in practice. Among emerging areas of interest for social scientists, the series presents machine learning methods, the use of new technologies for the collection of data and new techniques for assessing causality with experimental and quasi-experimental data.

Cambridge Elements

Quantitative and Computational Methods for the Social Sciences

Elements in the Series

Images as Data for Social Science Research: An Introduction to Convolutional Neural Nets for Image Classification
Nora Webb Williams, Andreu Casas and John D. Wilkerson

Target Estimation and Adjustment Weighting for Survey Nonresponse and Sampling Bias
Devin Caughey, Adam J. Berinsky, Sara Chatfield, Erin Hartman, Eric Schickler and Jasjeet S. Sekhon

Text Analysis in Python for Social Scientists: Discovery and Exploration
Dirk Hovy

Unsupervised Machine Learning for Clustering in Political and Social Research
Philip D. Waggoner

Using Shiny to Teach Econometric Models
Shawna K. Metzger

Modern Dimension Reduction
Philip D. Waggoner

Text Analysis in Python for Social Scientists: Prediction and Classification
Dirk Hovy

Interpreting Discrete Choice Models
Garrett Glasgow

Survival Analysis: A New Guide for Social Scientists
Alejandro Quiroz Flores

Adaptive Inventories: A Practical Guide for Applied Researchers
Jacob M. Montgomery and Erin L. Rossiter

A Practical Introduction to Regression Discontinuity Designs: Extensions
Matias D. Cattaneo, Nicolas Idrobo, and Rocío Titiunik

Bayesian Social Science Statistics: From the Very Beginning
Jeff Gill and Le Bao

A full series listing is available at: www.cambridge.org/QCMSS

Milton Keynes UK
Ingram Content Group UK Ltd.
UKHW031339281024
450123UK00020B/143